Innovation and Global Issues with Multidisciplinary Perspectives

Nurettin Bilici / Ragıp Pehlivanlı / Dilara Zorlutuna eds.

Innovation and Global Issues with Multidisciplinary Perspectives

Economics, Law and Social Sciences

Bibliographic Information published by the Deutsche Nationalbibliothek
The Deutsche Nationalbibliothek lists this publication in the Deutsche
Nationalbibliografie; detailed bibliographic data is available online at
http://dnb.d-nb.de.

Library of Congress Cataloging-in-Publication Data
A CIP catalog record for this book has been applied for at the Library of Congress.

InGlobe Academy

Cover image: https://www.pexels.com

ISBN 978-3-631-77488-5 (Print)
E-ISBN 978-3-631-78161-6 (E-PDF)
E-ISBN 978-3-631-78162-3 (EPUB)
E-ISBN 978-3-631-78163-0 (MOBI)
DOI 10.3726/b15274

© Peter Lang GmbH
Internationaler Verlag der Wissenschaften
Berlin 2018
All rights reserved.

Peter Lang – Berlin · Bern · Bruxelles · New York ·
Oxford · Warszawa · Wien

All parts of this publication are protected by copyright. Any
utilisation outside the strict limits of the copyright law, without
the permission of the publisher, is forbidden and liable to
prosecution. This applies in particular to reproductions,
translations, microfilming, and storage and processing in
electronic retrieval systems.

This publication has been peer reviewed.

www.peterlang.com

Contents

Hasan Bülent Kantarcı and Doğukan Salih Kutlutürk
Reflections of Externalities in the Public Economy 9

Özcan Öztürk
Impact of Microfinance on Small Enterprises in India 21

A. Öznur Ümit and Işıl Alkan
Do Credit Rating Agencies Predict or Deepen Financial Crises? 27

Muhammet Yunus Şişman
Volatility and Foreign Direct Investment in MENA Region: A Spatial
Panel Approach 45

Bilge Nur Öztürk and Tolga Öztürk
An Alternative Lifestyle Practice in a Globalizing World: Voluntary
Simplicity and Cittaslow 61

Serap Pelin Türkoğlu and Yasemin Hancıoğlu
Estimation of Countries' Development Status with Logistic Regression
Analysis 77

Hüseyin Yılmaz and Abdullah Elmas
The Study on Turkey's Demographic Window of Opportunity 91

Fatih Çağatay Cengiz
The Politics of Development of Turkey's Indigenous and National
(*Yerli ve Milli*) Defense Industry 99

Ferihan Polat and Ömer Ayna
Moral Behavior Types of Prospective Public Administrators: A
Case Study in the Department of Political Sciences and Public
Administration at Pamukkale University 113

Bilge Ünal
Cultural Dimensions in Migrant Literature Stereotyped Usage
in Authentic Stories 131

Hande Ünsal
Admission of Foreign Real Persons in Turkey 145

Osman Erdal Şahin
Planning and Spatial Regulation in İstanbul from Tanzimat
to the Republican Era 171

List of Contributors

Abdullah Elmas
Lec., Siirt University, Vocational School of Social Sciences, abdullahelmas@siirt.edu.tr

Bilge Nur Öztürk
Asst. Prof., Alanya Alaaddin Keykubat University, Faculty of Management, bilge.ozturk@alanya@edu.tr

Bilge Ünal
Lecturer, Bilecik Seyh Edebali University Rectorate, bilge.unal@bilecik.edu.tr

Doğukan Salih Kutlutürk
Doctoral Student, Kocaeli University, Institute of Social Sciences, dogukan_k@hotmail.com

Fatih Çağatay Cengiz
PhD, Res. Asst., Ondokuz Mayis University, Faculty of Economics and Administrative Sciences, cagatay.cengiz@omu.edu.tr

Ferihan Polat
Assoc. Prof. Dr., Pamukkale University, Faculty of Economics and Administrative Sciences, fyildirim@pau.edu.tr.

Hande Ünsal
PhD in Law, Asst. Prof., Ondokuz Mayıs University, Faculty of Economics and Administrative Sciences, hande.unsal@omu.edu.tr, handeunsal@gmail.com

Hasan Bülent Kantarcı
Assoc. Prof. Dr., Kocaeli University, Faculty of Economics and Administrative Sciences, hbkantar@kocaeli.edu.tr

Hüseyin Yılmaz
Doctoral Student, Gaziantep University, Institute of Social Sciences, huseyinyilmaz@siirt.edu.tr

Işıl Alkan
Asst. Prof., Ondokuz Mayıs University, Faculty of Economics and Administrative Sciences, isilalkan@omu.edu.tr

Muhammet Yunus Şişman
PhD, Dumlupinar University, Faculty of Economics and Administrative Sciences, myunus.sisman@dpu.edu.tr

Osman Erdal Şahin
Res. Assist., Uludağ University, Faculty of Economics and Administrative Sciences, osmanerdalsahin@gmail.com

Özcan Öztürk
Asst. Prof., Ataturk University, Faculty of Economics and Administrative Sciences, ozcan.ozturk@atauni.edu.tr

Ömer Ayna
Graduate Student, Pamukkale University, Institute of Social Sciences, Department of Political Science and Public Administration, omerayna29@gmail.com

A. Öznur Ümit
Assoc. Prof. Dr., Ondokuz Mayıs University, Faculty of Economics and Administrative Sciences, oumit@omu.edu.tr

Serap Pelin Türkoğlu
Res. Asst. Dr., Giresun University, Faculty of Economics and Administrative Sciences, serappelinozturk@hotmail.com

Tolga Öztürk
Asst. Prof., Alanya Alaaddin Keykubat University, Faculty of Management, tolga.ozturk@alanya@edu.tr

Yasemin Hancıoğlu
Asst. Prof., Ordu University, Ünye Faculty of Economics and Administrative Sciences, yaseminhancioglu@gmail.com

Hasan Bülent Kantarcı and Doğukan Salih Kutlutürk

Reflections of Externalities in the Public Economy

1 Introduction

It is possible to talk about the externality in the case that the actors in the economy or the companies have an indirect effect on the other individuals or firms and the actors who are under the influence do not bear the effect. If the actions and activities of the actors are indirectly beneficial to other actors, the validity of positive externality is mentioned here. For example, in the case of the establishment of a factory, the factory provides positive externality to the factory environment, such as the business opportunities provided by the factory, taxes paid to the state. However, if this indirect effect is harmful to the benefit rather than benefiting it, that is, if it is costing, there is the existence of negative externality. The most well-known example of negative externality is environmental pollution caused by producers in the production process (Stiglitz, 1994, p. 262).

Depending on the growth of the economies, the externalities reaching large dimensions have revealed the complexity of foreignness which is considered as a simple concept. Thus, there are many different separations in externalities. At first, it is only possible to talk about the externalities of the network, externalities in the education and health sector, and even environmental externalities, while talking about the existence of externalities in the production and distribution of public goods (Çetin, 2005; s. 144).

In the work of Adam Smith published in "The Rich of Nations", there are talks about social activities. Therefore, it is claimed that the concept of externality is the first time to reveal the author. Marshall then used these factories to explain the incremental turnover of firms, while exploring the costs they incur in the production of these factories, apart from the internal economies, in order to explain the economic growth performances and the productivity of the developed countries, especially in the UK.

1.1 Concept and Scope of Externality

The first theoretical origins of the concept of externalities were created by Marshall. (Yüksel & Kargı, 2010, p. 184). It was later tried to be mathematically expressed by Pigou and neo-classical economists. In this context,

$$FA = FA (X1, X2, ..., XN)$$

$$FB = FB (Y1, Y2, ... YN)$$

FA and FB consumers have utility functions, and these consumers benefit from consuming X and Y goods. In such a case, if the consumption activity of one of the consumers is effected by the utility of the other consumer, that is, if it is in the utility function,

$$FA = FA (X1, X2, ... Y1)$$

In such a case, there is talk of externality. In other words, it affects prosperity of product A consumed by consumer B (Önder, 2012, pp. 7–8).

The same situations can occur among producers. Externalities can then emerge in different forms. These producers are producers, producers are consumers, consumers are producers, and consumers are consumer goods (Giray, 2012, p. 13). In the externality of the producers, the production activities of a firm affect the production possibilities of another producer positively or negatively. For example, a decrease in the productivity of products produced by a manufacturer operating in the seafood sector resulting from the dropping of a company's wastes into the sea or to the sea.

When the benefits provided by the producers to the consumers are examined, technology confronts. Technologic products such as televisions, computers, and mobile phones have been made available to consumers and, over time, the cost of access to these inventions has been made easier by the manufacturers' technology development and use It is possible to realize the externality towards the producers from the consumers. For example, it is the benefit of the people who produce honey from the flowers that the people grow in their own gardens. A consumer's benefit to other consumers, for example, is to pollute the environment and create garbage. It is then possible to see the reflection path from one section to another of the types of externalities in the table below (Pehlivan, 2008, pp. 45–46). In the case of negative externalities, the output will be lower than the effective level if positive externality occurs, while the marginal social cost of the output is higher than the marginal specific cost (Bakırtaş, 2015, pp. 63–65).

Another distinction between externalities is the distinction between marginal and inframarginal externalities. According to this, there is a marginal externality if there is a positive or negative additional change in the welfare level of the other producer or consumer as a result of production or consumption activity. If the affected producer or consumer is not affected by the marginal changes in the activity that creates externalities, that is, if the marginal utility of the activity

Tab. 1: Types of Externality and Related Examples. Source: Made by Authors.

DIRECTION OF EXTERNALITY	PRODUCER		CONSUMER	
	POSITIVE	NEGATIVE	POSITIVE	NEGATIVE
PRODUCER	Beekeepers make a positive fruit tree	Damage to farmers due to waste of factory	Construction of new roads thanks to thermal power plant established	Established thermal power plant causing environmental pollution
CONSUMER	A person making a positive comment on a product	A person making a negative comment on a product	The beautiful flowers in the home garden and positive influences of passersby	Damage to the environment caused by gas from the auto exhaust

that caused the externalization is zero, then it would be possible to talk about the inframarginal externality (Yüksel & Kargı, 2010; s. 188).

A final distinction is the distinction between monetary and information externality. Monetary externality is the effect of a production or consumption activity through market-price transfer, through changes in prices. Technological externality is called information externalities, in which information generated by technological activities in which the firm or industry develops is of economic benefit to other firms or industries that are not common to the costs incurred when such activities are carried out (Türkcan & Kumral, 2013, p. 2) (Yüksel & Kargı, 2010, p. 189).

2 External Economies

Market disruptions are not the case in economies operating under perfect competition conditions. In other words, the markets where many buyers and sellers have transparent and full knowledge, and where these market actors buy and sell homogeneous goods, and there are no barriers to entering and exiting the market (Bulutay, 1982: pp. 146–147). It is not the case that external economies arise from the consumption of goods in the markets where all these features exist. That is, the consumption of private goods results in the actor benefiting only that consumer. However, the existence of such a market today is almost impossible. Other members of society may also benefit from such consumption and may be incurred.. In this case, deviations from the full competition market and market disruptions will arise, and external economies will result in inefficiency and deterioration in resource allocation (Orhan, 1984, p.72). Market failures usually arise in situations such as externalities, costly increases, natural monopolies, asymmetric information, and income inequality. These are leading factors in the growth of the size of macroeconomic imbalances and inefficiency in the markets.

It can only be insufficient to express external economies from the monetary side. If the movements of a firm affect the out-of-price conditions of other firms, they are mentioned in the existence of external economies. However, where market prices are not exogenous, the marginal social cost or marginal social benefit of goods and services cannot be fully expressed. While it is possible to determine in which direction the externalities form the economy and which parties influence the economy in its current form, it is difficult to pinpoint the benefits that the parties have from externalities and the costs they incur (Bakırtaş, 2015, pp. 61–62).

2.1 Consequences of Externality

The distribution of resources in the markets with externalities will deteriorate. The level of expenditure and production for externality will not reflect reality. The demand curve will show the marginal utility of the individual from the last unit of consumption, and the supply curve will show the marginal cost arising from the production of the last unit. The marginal cost at the intersection of the two curves will be equal to the marginal benefit. In the case of externality, the marginal private cost does not include environmental externalities, so the supply curve, which reflects only the specific costs, will become effective in the market, not the social cost.

In other words, the costs burdened on producers in the society will be reflected. In this case, where there is inefficiency in resource distribution, production will be reduced and production will be in the amount of balance, which is the point of intersection of the marginal social cost curve and the marginal utility curve. In such a case, it can be said that there is an over-production of goods which leads to negative externalities (Stiglitz, 1994, pp. 263–265).

In the same way that the source inefficiency is on the market, marginal social benefit is likewise lower than marginal private benefit. In such a case, the individual's return will be at the average rate of return, which is marginally higher (Stiglitz, 1994; s. 265).

3 Internalization and Compensation of Externalities

The problem of the inclusion of externalities in market prices in the economic discipline has led to many debates, and the problem has moved to the international dimension. Here two visions come out. The first argues that externalities need to be internalized by creating economic units, in which many consequences of economic activities remain within the unit, without requiring any state intervention. On the other hand, the market-price mechanism has failed to survive the externalities of the market. Because of this, active intervention of the government to the market has become compulsory. The state has taken this active intervention; taxes, subsidies, legal regulations-restrictions and production-taking methods, and fiscal policy instruments (Yüksel & Kargı, 2010, p. 191).

The internalization of externality is generally set in such a way that the marginal private benefit or cost of goods and services is equal to the social benefit or cost, while both approaches act on the same path. That is, in an economy where negative externalities exist, the method of internalization is carried out by adding marginal external costs to marginal private goods. Where positive externality applies, marginal private benefit to marginal external benefit in the internalization method should be added. The internalization method has important problems because the external benefit and external cost cannot be precisely determined (Mas-Colell, Whinston, & Green, 1995, s. 356).

3.1 Public Internalization Policies

Companies operating in the market do not take into consideration the factors that constitute external benefits and costs. Therefore, the state reminds the company of this situation with various measures and takes strict measures.

3.1.1 Pigou-Type Taxes

The main reason for resorting to taxation in the process of internalizing externalities is to try to remove the difference between social benefit and private benefit, social cost and custom cost. According to Pigou, in the presence of the negative external economy, the optimal condition requires a factor that causes negative externalities or a tax that is appropriate for the consumption or use of the goods (Kesbic, Baldemir, & İnci, 2010, pp. 131–132).

Tab. 2: Classification of Public Policy Instruments Against Externalities. Source: Made by Authors.

POLICIES	DIRECT TOOLS	INDIRECT TOOLS
Market-focused interventions	Government collected fees	State support and tax

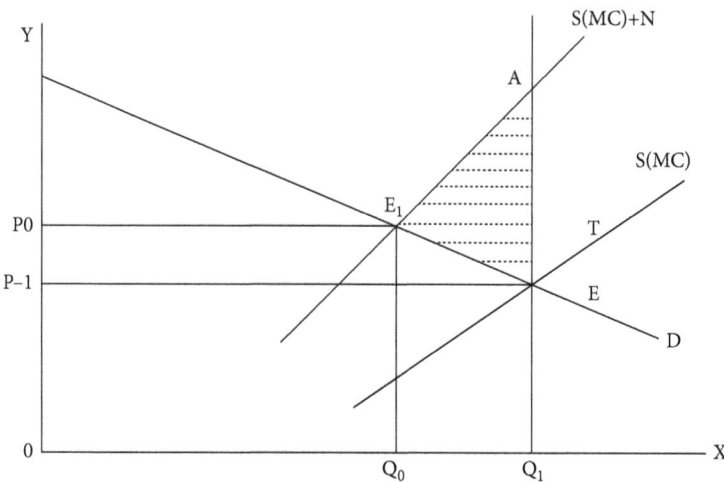

Fig. 1: Regulatory Taxes and Outside Costs. Source: Kesbiç, Baldemir, & İnci, 2010, p. 132

A tax of type Pigou is a tax on a unit of negative externality that is derived from each unit of the output of the same amount of marginal damage. The vertical distance between marginal social cost and marginal private cost equals marginal damage. The producer has to make payments to both suppliers and taxpayers for each unit production. Geometrically, a person's new marginal cost curve is found by adding marginal special finance at every level of output.

It is possible to produce at the highest price point where the marginal cost is equal to the marginal benefit. This level of production occurs at the intersection of marginal utility. The economic actors will also add to the cost of externalities, depending on the tax. The applied tax causes a certain increase in income for each unit of product being produced.

3.1.2 Plott's Regulatory and Tax Approach

Plott considered Pigou-type taxes as regulatory taxes and analyzed the effects of regulatory taxes on goods that caused negative externality. Plott argues that regulatory tax should be used in the face of negative externalities. On the contrary, Buchanan stated that the applied regulatory taxes could cause welfare loss in the markets (Kesbiç, Baldemir, & İnci, 2010, p. 132).

In Fig. 1, the S (MC) + N curve represents negative external economies. In this case, the marginal cost curve of the firm shifts upward and the firm balance occurs at the E1 point. This point is far from optimal in terms of society. Because, at this point, the company is charging external costs as much as collecting EA. Here, using regulatory taxes, the community can come back to the optimum point. In the example, a regulatory tax is applied up to EA, providing net social benefit as much as E-E1- A.

3.1.3 Fees

The fees are an unexpected source of income, unlike the tax received by the state. Accordingly, the actors in the market are collecting fees from the people who cause negative externalities. Thus, the negative externality is tried to be minimized by the public activities carried out with collected fees.

4 Market Solutions Against Externalities

4.1 Coase Teoremi

In the case of externalities, it is called Coase Theorem to propose that some of the actors in the market intervene in a way to internalize externality and achieve some efficiency. Coase theorem does not include any active intervention of the state on the market. According to this, when the problem of inefficiency in the market arises, the parties bargain with each other. In this way, they play a role in internalizing the problem of externality.

In order for the Coase Theory to be successful, ownership rights must be established to market actors. It is thus possible to create a market for externalities and to remove this from market failure. Coase Theory focuses on transaction costs termination (Kesbic, Baldemir, & İnci, 2010, p. 127); the possibility of negotiating between decision makers is higher and that this fact provides the optimality if the property rights are traded without obstacles and the transaction costs are low. Coase argued that a mechanism could be realized that could provide the optimal distribution of resources in the economic system where there are perfectly competitive conditions, even in the presence of significant externalities.

The basic assumptions for Coase's work are listed as follows (Bastürk, 2014, p. 147):

- The cost system does not cost
- No disruptions in the property system
- The absence of asymmetric information
- Transaction costs should be close to zero or near zero.

Although the Coase theory cannot be fully proved, it cannot be refuted. That is to say, in the absence of transaction costs, the distribution of activity and property rights can be achieved independently as a result of agreements between actors. In the real world, however, the relations that this theory holds can be seen in a very small area. The main reason for this is the increase in transaction costs due to the increase of actors, the problem of asymmetric information, the cost of exploiting property rights, and the inability to buy and sell property rights to the value judgments of the society (Çetin, 2005).

4.2 Hicks–Kaldor Criterion

The Hicks–Kaldor Criterion is also called the Compensation Solution or the Welcoming Loss Initiative. The main aim of this criterion is to broaden the scope of the application of the Pareto criteria without state intervention and to bring the principle of compensation by eliminating uncertainty. If there is no possibility of raising the prosperity of another actor without reducing the prosperity of an actor in society, the prosperity of society according to Pareto is the optimal level. When Kaldor and Hicks are in such a state of equilibrium (state B) with Pareto optima (state B), if the gain of those beneficial in changing this state of equilibrium is higher than the losses of the harmful ones, the increase in social wealth is a result of such a change of balance (Kesbiç, Baldemir, & İnci, 2010, p. 129). So in this case, if the actors that are profitable compensate the harm of the actors who are harmful, if they can still make profits, the social prosperity also increases.

In Fig. 2, when the TTI is moved from the D point between the welfare limit and the origin to the E point on the welfare limit, there is a decrease in the satisfaction level of the individual no. 1 and an increase in the individual no. 2. In order for the E point to fulfill the said parallel, the individual no. 2 must compensate the actor of the no. 1 actor so that the two actors may be able to increase their initial welfare levels.

4.3 Scitovsky Approach: Bargaining Criteria

This theory, which is presented as an alternative to the Kaldor–Hicks approach, is also called "bargaining criterion". The theory is from Tibor Scitovsky. If, according to this approach, an economic unit imposes an external cost on another economic unit as a result of the consumption activity on the production side, then the external cost unit must engage in bargaining to limit the activity of the economic unit that is causing it. This bargaining can be in the form of reconciliation between the two sides (Öz & Buyrukoğlu, 2012; s. 8).

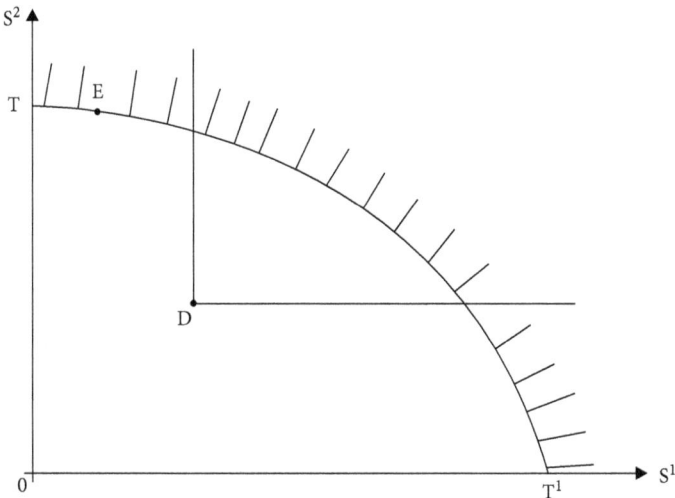

Fig. 2: Hicks – Increase in Social Welfare According to the Caldor Measure.
Source: Kesbiç, Baldemir, & İnci, 2010, p. 130

4.4 Comparison of Public Economics and Market Solutions in Internalizing Externalities

As mentioned in Coase theory, market solutions are insufficient at some points in the internalization of externalities. Especially in property rights, the system is not developed in every country and economy, and the transaction costs between actors are not always low, weakening the validity of this theory. On the Hicks-Kaldor scale, the agreement of market actors and compensation for damages is a temporal and materially separate cost element and also does not take into consideration the elements of income distribution. It is theoretically possible that the parties can negotiate without public intervention. However, market instruments are sensitive to technological innovations and lead to these innovations.

In the case of public economic instruments, however, different application difficulties can be encountered. For example, if the most optimal level of Pigou-type tax is determined, marginal loss or benefit should be at an equal level between the actors. However, it is not always possible to calculate the monetary values in the market of marginal social and private harm.

5 Result

Externality is defined as both the producers and the producers of goods and services, the benefits and losses that affect third parties, and the inputs that do not have price barriers in the market. Many measures have been developed to internalize externalities in order to minimize these benefits or losses, but most of these criteria do not account for most of the economic activities in the real world. One of the main reasons for this is the high transaction costs. On the other hand, attempting to avoid activities that cause externalities as a result of public intervention with public means such as taxes and subsidies can incur high costs to the state. Therefore, this practice is not always valid due to the political process. The level of taxation and subsidy is not always certain. Because, in the market, the social cost and the availability of benefits are not always fully identifiable. Hence, a mixed intervention should be made in externalities as a combination of both market instruments and public means. In order for this hybrid application to be valid, the issues such as income distribution justice, public revenues, and resource budget should be carefully examined.

References

Bakırtaş, I. (2015). Dışsallıklar Sorununun İçselleştirilmesinde Düzenleyici Vergiler ve Sübvansiyonların Etkinliği: Analitik Bir Yaklaşım. Dumlupınar Üniversitesi Sosyal Bilimler Dergisi, No;7, p.57-72 .

Baştürk, M. F. (2014). Mülkiyet Problemi, Dışsallıklar ve Coasean Çözüm. Yönetim ve Ekonomi, 21,1, 143–154.

Bulutay, T. (1982). Dışsal Ekonomilerin Anlamı ve Önemi Üzerine. Ankara Üniversitesi SBF Dergisi, 37, 1, 141–156.

Çetin, T. (2005). Çevresel Dışsallıklar ve İçselleştirme Yöntemleri. Gazi Üniversitesi İktisadi ve İdari Bilimler Fakültesi Dergisi, 7, 3, 143–166.

Giray, F. (2012). Bilgi Ekonomisinde Entelektürel Mülkiyet Hakları Üzerinde Küreselleşmenin Etkileri ve Harmonizasyon Sorunu. İstanbul: Bilgesam Yayınları.

Kesbiç, C. Y., Baldemir, E., & İnci, M. (2010). Dışsallıkların Ekonomi Üzerindeki Etkileri ve İçselleştirilmesine İlişkin Teorik Yaklaşımlar - Çözüm Önerileri: Yatağan Termik Santrali Analizi. Yönetim ve Ekonomi Araştırmaları Dergisi, No: 14, 123–138.

Mas-Colell, A., Whinston, M. D., & Green, J. R. (1995). Microeconomic Theory. New York: Oxford University Press.

Önder, İ. (2012). Pareto Dengeleri Açısından Bölünmezlik ve Dışsallık Kavramları. Maliye Araştırma Merkezi Konferansları, 0, 23, 1-12.

Orhan, Ş. (1984). Kamu Ekonomisi. İstanbul: Okan Yayıncılık.

Öz, E., & Buyrukoğlu, S. (2012). Negatif Dışsallıkların Önlenmesinde Çevresel Vergiler: Türkiye ve OECD Ülkeleri Karşılaştırması. TİSK Akademi, 14.

Pehlivan, O. (2008). Kamu Maliyesi. Trabzon: Derya Kitabevi.

Stiglitz, J. E. (1994). Kamu Kesimi Ekonomisi (Batırel, Ö. F. Çev.). İstanbul: Marmara Üniversitesi Yayınları.

Türkcan, B., & Kumral, N. (2013). Yüksek Teknolojili Endüstrilerde Bilgi Dışsallıkları: İzmir Örneği. İzmir: Working Papers in Economics.

Yüksel, C., & Kargı, V. (2010). Çevresel Dışsallıklarda Kamu Ekonomisi Çözümleri. Maliye Dergisi, No: 159, 183-202.

Özcan Öztürk

Impact of Microfinance on Small Enterprises in India

1 Introduction

One of the issues in the field of finance is the lack of access to credit by the poor. Across the world, particularly in the developing countries, there is a consensus that microfinance, small long-term loans on easy terms, has the potential to alleviate poverty in rural areas and empower small businesses and the poor, especially women. Conventional banks are reluctant to lend money to the poor as the poor are unable to offer collaterals or any form of security. Micro-finance institutions (MFI), however, do not require collateral since they offer loans on easier terms and conditions. A well-known example of MFI is the Grameen Bank established by Nobel Peace prize recipient, Muhammad Yunus, in Bangladesh in the early 1970s. Since then, the Grameen Bank model has been adopted by many countries.

Microfinance is perceived as an important means for poverty reduction in India as well. It was first initiated by National Bank for Agriculture and Rural Development (NABARD) in collaboration with banks and NGOs to lend the rural population, known as Self-Help Group (SHG) program in the late 1980s. Usually NGOs provide SHG access to funds through banks. There were also state-run SHG programs. Thus, initially microfinance was dominated by government. In the 1990s, private sector-run MFIs entered microfinance market. Since then, in India there have been two channels providing microfinance to the rural population: MFIs and SHG bank linkage model.

Due to potential demand in the sector and high re-payment rate, the microfinance sector witnessed high growth rate during the period 2006–2010. However, the sector faced a crisis in 2011 due to regulatory uncertainty. In the state of Andhra Pradesh, state government intervened the market blaming that MFIs are the main cause of more than 200 suicides taken place due to over indebtedness. As a result, the sector witnessed funding constraints because of uncertainty which led the "microfinance crisis" in India. After the crisis in 2011, the central government regulated the market to revive it and bring it back on track.

Several studies have done impact analysis of MFIs in India. Imai et. al. (2010) examine whether household access to microfinance reduces poverty. Using national household data and treatment effect model, they find that loans for productive purposes were more effective in rural area than urban areas for poverty reduction. A vast majority of the studies has focused on the effect of MFIs on women empowerment. Sarumathi and Moha (2011) study the effectiveness of microfinance on women empowerment. They find a positive correlation between microfinance access and improvement in literacy, awareness for children education and reduction in poverty level. Aruna and Jyothirmayi (2011) analyze impact of microcredits on women empowerment and find microfinance has profound impact on the economic status, decision-making power of women.

Although there is a relatively large literature on impact of MFIs in individual level, to my knowledge, there is no impact analysis done in firm level. This study aims to fill this gap by providing evidence whether MFIs have improved small firms' access to credits or not.

The literature of MFIs' impact on small enterprises dates back to Hartarska and Nadolnyak (2008) who examine whether microfinance institutions improved access to credit for small enterprises in Bosnia and Herzergovina using the financing constraints approach. They find that MFIs alleviated microbusinesses' financing constraints. Abiola (2011) uses financing constraints approach to study whether MFIs has improved access to credit for small firms in Nigeria, and finds positive impact of MFIs in alleviating microbusiness' financial constraints.

Hartarska, Nadolnyak and McAdams (2013) study the effectiveness of MFIs on availability of credit to entrepreneurs in Eastern Europe and Central Asia using financing constraint approach. They compared cities with MFIs with the ones without MFIs and find that MFIs have improved credit access of small businesses. Using the same approach, Quaye and Hartarska (2016) examine the impact of MFI on micro- and small enterprises in Ghana. Results show that microfinance sector has alleviated financing constraint in Ghana.

2 Methodology

This study adopts the financing constraints approach used by Quaye and Hartarska (2016). The approach was first used by Fazzari et al., (1988) to test if there is a difference in sensitivity of investment to internal funds in enterprises that are characterized constrained as opposed to unconstrained ones. Hartarska and Nadolnyak (2008) extend the financing constraint approach to evaluate impact of microfinance industry on firms' access to credit. The authors create two sub-samples: one sample represents the microenterprises in cities with MFIs (treatment) and the other one without MFIs (control group). For each sub-group, a reduced form investment equation such as the one in equation (1) is estimated. If two groups indicate statistically different sensitivity to internal funds, one group of enterprise is more credit constrained. In other words, if investment in firms in cities with MFIs is less sensitive to internal funds as compared to cities without MFIs, one can argue that MFIs alleviate financing constraints.

3 Data and Estimation

This paper uses data from the World Bank Business Environment and Enterprise Performance Survey (BEEPS) collected in India between June 2013 and December 2014. Thus, one year of cross-sectional data is used for the analysis. The survey includes firms from different sectors including manufacturing, trade and services and from three different sizes of firms including small (with employees between 5 and 19) and medium (between 20 and 99) and large firms (with more than 100 employees). We analyze the small enterprises. We divide the sample into two sub-groups based on firms' self-reporting to create control and treatment groups. The firms that report "access to finance" as "little" or "no obstacle" are classified as unconstrained while the enterprises that choose a "moderate", "major" or "very severe obstacles" are classified as constrained. We have a final sample of 273 small enterprises with 114 constrained enterprises and 159 unconstrained firms. (Tab. 1a and Tab. 1b)

The model (1) is estimated for firms that have credit access and the ones that do not have access to credit.

Tab. 1a: Descriptive statistics for constrained enterprises in India.

Variable	Description	Type	Sample	Mean	Min	Max
Investment	Amount of investment made previous year	Continuous	114	959719.30	0	7800000
Cash Flow	Total revenue – total cost	Continuous	104	17363461.21	184000.00	209160000
Ent. Age	Enterprise age	Continuous	114	17	1	66
Perm employ	Permanent employees	Continuous	114	12	5	19
Categorical variable				**Categories**	**Percent**	
Investment opportunity	Part-time worker: yes=1, no=0	Dummy	114	0 1	51.75 48.25	

Tab. 1b: Descriptive statistics for unconstrained enterprises in India159Dummy.

Variable	Description	Type	Sample	Mean	Min	Max
Investment	Amount of investment made previous year	Continuous	159	835918.24	0	7800000
Cash Flow	Total revenue – total cost	Continuous	152	20989819.46	184000.00	209160000
Ent. Age	Enterpise age	Continuous	158	19	2	84
Perm employ	Permanent employees	Continuous	159	22	5	19
Categorical variable				Categories	Percent	
Investment opportunity	Part-time worker	Dummy	159	0 1	57.86 42.14	

$$Inv = \beta_0 + \beta_1 IF + \beta_2 IO + \beta_{3-i} Z + \varepsilon \qquad (1)$$

Where Inv is the amount of investment made in previous year, IF is internal fund (cash flow) available which is obtained by subtracting total cost from total revenue. IO is investment opportunity. We use hiring part-time worker as proxy for investment opportunity. Including investment opportunity is particularly important since enterprises that do not use investment opportunity might not invest even if they have capital to do so. Z is a vector of other variables about the enterprises the affect investment. Finally, ε is iid error term with zero mean and constant variance. From the specification of the model (1), a statistically significant difference in β_1 between two subsample suggests existence of impact. In other words, investments in small firms having access to microfinance are less sensitive to internal funds than the ones that do not have access to microfinance. Thus, one can infer that MFIs alleviate financing constraints of small enterprises in India.

4 Results

Tab. 2 presents Tobit regression results for both constrained and unconstrained small enterprises. We take log of investment and cash flow to correct violation of

Tab. 2: Tobit model parameter estimation results.

Investment	Unconstrained	Constrained
Cash Flow	0.2722*	0.4474*
	(0.0706)	(0.1042)
Inv Opportunity	-0.02663	0.6476*
	(0.025)	(0.242)
Enterprise age	-0.0031	-0.0101
	(0.0075)	(0.0090)
Perm employees	0.0093**	-0.0053
	(0.0027)	(0.0137)
Log likelihood	-250.343	-166.931
Observations	148	103
Intercept	8.4979*	5.9427*

Standard error in parenthesis. * and ** are 1 % and 10 % significance levels.

constant variance assumption. As seen in the table, the estimated parameter of cash flow for unconstrained firms is 0.2722 while it is 0.4474 for unconstrained firms. T-test results show the difference between two coefficients is statistically significant at 5 percent level. This indicates that the constrained enterprises are more dependent on internal funds than the unconstrained ones. As mentioned earlier, this suggests that having access to microcredits has alleviated financial constraint faced by small enterprises in India. As for investment opportunity, it is statistically significant suggesting that the constrained enterprises that have investment opportunity are able to make 65 percent more investment as opposed to the ones that do not have investment opportunity (part-time employees). Statistical significance of the variable of permanent employee for unconstrained enterprises suggests that an additional permanent employee has positive impact on investment for unconstrained firms. The rest of the variable do not have statistically significant effects on investments for both constrained and unconstrained firms.

5 Conclusion

This study examines the impact of microfinance institutions on small enterprises' access to credit in India. Using financing constraint approach and one year of data provided by Business Environment and Enterprise Performance Survey, we find that microfinance institutions have improved access to credit for small firms in India since the enterprises that have access to microfinance (unconstrained firms) were less sensitive to availability of internal funds than constrained enterprises that do not have access to microfinance. This study adds literature of another case where small enterprises' financing constraint can be alleviated by microfinance institutions.

References

Abiola B. (2011). "Impact Analysis of Microfinance in Nigeria," *International Journal of Economics and Finance* 3 (4), pp. 217

Aruna, M., Jyothirmayi M.R. (2011). "The role of microfinance in women empowerment: A study on the SHG bank linkage program in Hyderabad (Andhra Pradesh)." *Indian Journal of Commerce & Management Studies ISSN* 2229:5674.

Fazzari, S.M., Hubbard R.G., Petersen B.C., Blinder A.S., Poterba J.M. (1988). "Financing constraints and corporate investment." Brainard W.C. and Perry G.L.(Eds.), *Brookings Papers on Economic Activity* 1988(1):141–206.

Hartarska, V., Nadolnyak D. (2008). "An impact analysis of microfinance in Bosnia and Herzegovina." *World Development* 36(12):2605–2619.

Hartarska, V., Nadolnyak D., McAdams T. (2013). "Microfinance and microenterprises' financing constraints in Eastern Europe and Central Asia." In *Microfinance in Developing Countries*. Springer, pp. 22–35.

Imai, K.S., T. Arun, Annim S.K. (2010). "Microfinance and household poverty reduction: new evidence from India." *World Development* 38(12):1760–1774.

Quaye, F.M., Hartarska V. (2016). "Investment impact of microfinance credit in Ghana." *International Journal of Economics and Finance* 8(3):137.

Sarumathi, S., Mohan K. (2011). "Role of micro finance in women's empowerment (an empirical study in Pondicherry region rural SHG's)." *Journal of Management and Science* 1(1):1–10.

A. Öznur Ümit and Işıl Alkan

Do Credit Rating Agencies Predict or Deepen Financial Crises?

1 Introduction

Credit rating agencies (CRAs) are seen as important institutions for solving the asymmetric information problem between issuer and investor in the process of globalization in which rapid flow of funds and international trade volume is increasing. The rating activity has become important in attracting foreign investments to the country and access to foreign funds has become difficult for unrated companies. In this regard, the ratings of the stated institutions have become a kind of "reputation" in international markets. On the other hand, whether or not the relevant institutions have realistic grades and predict the financial crises that need to be foreseen is a separate debate. Furthermore, the fact that these institutions do not deepen and deepen the financial crises is frequently questioned today. The purpose of this study is to give brief information about the CRA, to question the ability of these institutions to predict the financial crises that occur on a global scale, and to examine whether relevant institutions deepen or do not deepen crises.

In this framework, firstly, the definition and functions of the credit rating will be discussed. In the sequel, the short history of rating activity, major CRAs, and rating systems will be explained. Afterwards, their efficiencies in forecasting or deepening financial crises will be questioned within the context of Southeast Asia, Russia, Brazil, Turkey, Dotcom crises, Enron and Parmalat scandals, American mortgage crisis, bankruptancy of Lehman Brothers, and Greece debt crisis.

2 Definitions and Functions of Credit Rating

A rating by its most general definition is an activity that measures the ability of a debtor to pay his debt (principal and interest liability) and desire for payment with a number of financial analyses (Kılıç, 1989: 8). According to the Regulation of the Banking Regulation and Supervision Agency, the rating activity is determining the credit worthiness of the customers and thus the rating grade. The determination of creditworthiness is the independent, impartial and fair evaluation, classification, and appropriate grading of the risk that the customer will not be able to meet the principal, interest, and similar liabilities of the capital market instruments representing the repayment power or debt of the customer (Resmi Gazete, 2012).

According to the definition of the Capital Markets Board, credit rating is "an independent, impartial and fair evaluation and classification of rating institutions' ability to meet the risks, repayments of the enterprises or the maturity of capital, interest and similar liabilities of the capital market instruments representing debt" (SPK, 2018). As determined in these definitions, the rating is an assessment of the risk that the investor will undertake. The rating aims to reveal the ability of the issuer to meet its debt solvency, financial structure, and liabilities on time. Fund providers seek to measure the creditworthiness of the country or institution that requests funds. In this context, there is an adverse relationship between credit ratings and risk, likewise, a higher rating implies the ability to meet obligations, while a poor rating indicates poor financial structure and inadequate solvency (Hamzo, 2007: 29). Briefly, the rating activity is a kind of risk analysis, the borrowers are analyzed according to various criteria and graded, and grades assigned to borrowers are indicative of full and timely payment of debts. These indicators are letters, numbers, or various symbols made up of their combination (Özdinç, 1999: 7).

The increasing volume of international trade volume and fund flows in the globalization process has enabled the role of credit rating agencies, and related institutions have become a major source of information for investors in many parts of the world. In particular, investors who want to invest in a foreign country have gained the opportunity to assess what scale and what kind of risk it takes by looking at the rating grades, in this framework, the involvement of the companies which are not subject to the rating activity to the global markets have become difficult, and the credit ratings of countries and companies have become a kind of "reputation" in the international fund market. However, it should be underlined that the role of CRAs in financial markets has increased considerably, especially after the development of Basel 2 capital standards. These organizations have started to be considered as very important institutions because they are the corporations that solve the problem of information asymmetry between the issuer and the investor. High credit ratings have become a prerequisite for funding at favorable interest rates and currencies in international markets, so issuers with low credit ratings can get high risk premiums and raise funds at high interest rates compared to issuers with high ratings (Elkhoury, 2008: 1). In this context, the three main functions of CRA can be mentioned. The first is to measure the credit risk of the issuer, the second is to provide the possibility of comparison, and the third is to develop a common standard. The first function is very important in terms of removing information asymmetry between the issuer and the investor.

The second function allows the investor to compare the risks of all the investment instruments in the process of creating the portfolio. The third function attaches importance to the development of a common standard, language capability in the sense of credit risk to market participants (OECD, 2010).

Credit rating agencies rate not only institutions and securities but also countries and local governments. In this context, it should be emphasized that country ratings are highly determinative in terms of borrowing costs at international markets. Definitely, the credit rating of an individual country is the key determinant for the interest rates that the country will face in international financial markets. High interest rates have the effect of increasing the cost of borrowing. On the other hand, the given rating of a country also has a restrictive effect on ratings given to the institutions in the same country (Afonso et. al., 2007: 7). In other words, credit ratings of institutions operating in a country with a low credit rating will not be high (due to country risk). For this reason, credit ratings of countries are not only connected to the public sector, but also closely related to private sector companies.

There are two types of rating, long and short, according to the maturity. A long-term rating is a measure of the ability of a country or company to meet its long-term obligations on the basis of key economic, cyclical, technological, and legal regulations. A short-term rating is a measure of the ability of a country or financial institution to access money and capital markets, obtain liquidity and capital, while meeting short-term obligations for one year. There are two types of ratings depending on their type: foreign currency and local currency. Moreover, the company rating in foreign currency cannot exceed the country rating; in short, the country rating is a "ceiling" (Akbulak, 2012: 172).

3 Brief History of Credit Rating Activity and Key Rating Agencies, Rating Systems

The economic deterioration and high inflation in the United States between 1833 and 1837 concluded with the economic crisis, and in this period investors and traders were in great need to have knowledge of the credit worthiness of partnerships or companies they were doing business with. In response to this need, Lewis Tappan, one of the investors who were adversely affected by the crisis in 1841, tried to establish a database which could monitor the financial health of the companies, evaluate the financial structures, and reveal creditworthiness and thus Mercantile Agency was founded (Destraz & Lahaye, 2012: 1). Over time, this business became a profitable activity for Tappan, and other organizations followed Tappan's path. In 1890, Poor's Publishing Company, the predecessor of Standard & Poor's (S&P), issued a guide called "Poor's Book", which included extensive data on railway bills and business analysis. All of these originate the credit rating industry, which sprang up at the beginning of the 1900s. John Moody, one of Wall Street's financial analysts, then put the businesses under the scope that were in commercial activity and found significant problems about those businesses. He published his book consisting the analyses in 1909, the book which was sold throughout the United States turned out to be an important financial instrument revealing the credit quality of the companies and led to the development of the credit rating industry. In 1914, Moody's was established and the rating department of the founding department began to work effectively in 1922. Following these developments, Poor's Publishing Company, the Standard Statistic Company, and the Fitch Publishing Company started to operate respectively in 1916, 1922, and 1924 (Pascalis, 2017: 16–17).

Despite the increasing number of CRAs today, the three major companies in the world have 93 % of the market. These companies are Standard & Poor's, Moody's and Fitch rating agencies. Standard & Poor's is the dominant company of the market with 45 % market share. Moody's is the second largest company with a share of 31.29 %, and Fitch is the third largest company with a share of 16.56 % (ESMA, 2016: 6). Standard & Poor's, the market dominant, is an establishment in central America with analysts and economists in 28 countries which has been rating companies, financial institutions, and securities over 1 million for more than 100 years (S&P, 2018). Moody's, the second largest company of the market, is an establishment in central America with an employment network of approximately 12,000 people operating in 42 countries (MOODY'S, 2018). Fitch, the third largest rating agency, operates in more than 30 countries and has two separate centers in London and New York (FITCH, 2018).

The credit rating agencies express the result of the rating activity by letters, numbers, symbols, or a mixture thereof as explained before. On the other hand, the grading systems of the related institutions differ from each other, but they are also similar in certain respects. For example, level A shows a high overall rating for most rating agencies, but the highest rating is AAA for Standard & Poor's and Fitch, and Aaa for Moody's. The BBB or Baa2 levels indicate the levels below the median and generally describe securities, financial institutions, or countries with a speculative value below the BBB- or Baa3 level, where the investment is at risk (Tab. 1).

4 Credit Rating Agencies and Financial Crises

This section scrutinizes the forecasting ability of credit rating agencies (CRA) in the face of various financial crises in the world and deals with their role in the

Tab. 1: Rating system of main credit rating agencies and description. **Source:** Karagöl & Mıhçıkour, 2012: 16; S&P, Moody's and Fitch.

Credit Rating Agencies Rating System				
Standard & Poor's	Fitch	Moody's	Description of Rating	
AAA	AAA	Aaa	Highest grade	**Investable degree**
AA+	AA+	Aa1		
AA	AA	Aa2	High grade	
AA-	AA-	Aa3		
A+	A+	A1		
A	A	A2	Medium grade	
A-	A-	A3		
BBB+	BBB+	Baa1		
BBB	BBB	Baa2	Lower medium grade	
BBB-	BBB-	Baa3		
BB+	BB+	Ba1		**Speculative degree**
BB	BB	Ba2	Non investment grade	
BB-	BB-	Ba3		
B+	B+	B1	Speculative	
B	B	B2	Highly speculative	
B-	B-	B3		
CCC+	CCC	Caa		
CCC	CC	Caa3	Substantial risks	
CC	C	Ca	Extremely speculative	
D	DDD	D	In default	**Default**
	DD			
	D			

economy concerning relevant crises. To this end, it discusses Southeast Asian, Russian, Brazilian and Turkish crises, the dot-com crisis, the Enron and Parmalat scandals, the United States subprime mortgage crisis, the bankruptcy of Lehman Brothers, and the Greek debt crisis, respectively.

4.1 Southeast Asian crisis

The Southeast Asian crisis started in July 1997 with the devaluation of the Thai baht. The crisis starting in Thailand rapidly spread throughout the Philippines, Malaysia, Indonesia, and South Korea and turned into a regional crisis. The excessive amount of speculative foreign capital inflow to the region and its fragile financial structure were the two most important underlying reasons for the crisis (Engin, 2007: 43). To put it differently, the most important cause of the crisis was the lack of a sound financial system in the countries in the region and, by extension, banks' wrong loan and credit policies. In the aftermath of the financial crisis, the crisis-ridden economies witnessed increased unemployment, a sharp decline in international credit flows, the bankruptcy of companies and banks, a rapid increase in interest rates in national and international markets, and a decline in world trade volume (Turan, 2011: 61).

Additionally, the Southeast Asian crisis led to growing concerns about the credit rating of CRAs. A number of shortcomings were noticed in the Asian financial system in the pre-crisis period; however, CRAs downgraded their credit ratings either before or after the crisis began (Çevik Tekin, 2016: 193). In fact, in the aftermath of the crisis, Standard & Poor's downgraded its credit rating for Indonesia and South Korea by eight notches, for Malaysia by five notches, and for Thailand by four notches. While S&P rated Thai government bonds as A on July 1, 1997, it downgraded its credit rating to BBB- after the crisis. Likewise, in the aftermath of the crisis, S&P downgraded its credit ratings in July 1997 for Indonesia from BBB to CCC+, for South Korea from AA- to BB+, and for Malaysia from A+ to BBB-. The financial markets, on the other hand, showed a strong reaction to this situation (Kräussl, 2000: 5).

Market participants criticized CRAs for not forecasting crises, arguing that their credit rating was rather a response to the negative situation in the country. Participants' another criticism of CRAs was that downgrading the credit ratings after the financial crisis started increased existing instability in the country. The reason for these criticisms was, as seen in Tab. 2, a Standard & Poor's downgrade of Indonesia's and South Korea's ratings by eight notches, Malaysia's rating by five notches, and Thailand's rating by four notches (Kräussl, 2000: 5). This situation indicates that CRAs downgraded their credit ratings by multiple notches at once in the aftermath of the financial crisis, CRAs were unable to make sound analyzes, their ratings were not reliable, and therefore they did not properly fulfill their functions.

Tab. 2: Standard & Poor's changing ratings during the Southeast Asian crisis. **Source:** Kräussl, 2000: 5.

Countries	November 30, 1998	July 1, 1997
Indonesia	BBB	CCC+
South Korea	AA-	BB+
Malaysia	A+	BBB-
Thailand	A	BBB-

4.2 Russian Crisis

With the dissolution of the Soviet Union in 1991, Russia shifting from a planned economy to a free market economy made the national currency ruble circulate in the international market, thereby making the Russian economy more vulnerable to risks and setting the stage for the financial crisis. Additionally, two main reasons triggered the outbreak of the Russian crisis. First, the fall in oil prices driven by the Asian crisis led to a drop in Russia's most important export revenues, thereby causing a problem of current account deficit in Russia. Second, Russia's high credit scores in the relevant years caused the government to borrow more short-term debts. To put it differently, the decline in oil prices reduced Russia's foreign exchange revenues. Therefore, Russia had difficulty paying back its short-term debts (Oktar & Yüksel, 2015: 331). S&P declared Russia's credit rating to be B+ on June 9, 1998 but downgraded it to B- on August 13, 1998 and to CCC on August 17, 1998. Due to similar reports of other CRAs, on August 17, 1998, the Russian government devalued the ruble, declared a moratorium, and defaulted on its debt (Toraman & Yürük, 2014: 141). (Tab. 3 for details)

Tab. 3: Standard & Poor's and Moody's credit ratings of Russia (January 1997–December 2000). **Source:** Kräussl, 2003: 45.

Date	S&P	Moody's
1.1.1997	BB–(N)	Ba2 (N)
19.12.1997	BB–(O–)	
3.2.1998		Ba2 (CW–)
11.3.1998		Ba3 (N)
27.5.1998	BB–(CW–)	
29.5.1998		B1 (N)
9.6.1998	B+ (N)	
13.8.1998	B–(O–)	B2 (N)
17.8.1998	CCC (O–)	
21.8.1998		B3 (N)
16.9.1998	CCC–(O–)	
27.1.1999	SD	
10.4.2000		B3 (O+)
23.8.2000		B3 (CW+)
13.11.2000		B2 (N)
8.12.2000	B–(N)	

4.3 Brazilian Crisis

With the Real Plan (a Foreign Exchange-based Stabilization Program) put into force in 1994, Brazil reduced inflation and had a successful growth performance from 1994 to 1998. However, Brazil failed to finance its budget and current account deficits. Thus, its budget deficit rose to 8 % and its current account deficit to 4.5 % (IMF WEO, May1998: 28). Russia's declaration of default on its foreign debt in August 1998 also caused investors to re-evaluate their positions in emerging markets. Thus, Brazil attracted investors' interest with its high budget and current account deficits. Eventually, Brazil panicked by the Russian crisis financed its current account deficits with short-term capital flows and ended with a capital outflow of 30 billion dollars within three months following the August of 1998. In line with all these incidents, the Brazilian real was devalued by 9 % and the fixed-exchange-rate system was abandoned on January 13, 1999 (Yavaş, 2007: 90). Due to the crisis in the country, S&P downgraded Brazil's credit rating by one notch from BB- declared on September 10, 1998 to B+ on January 14, 1999 (Kräussl, 2003: 41).

4.4 Turkish Crisis

The financial liberalization process in Turkey accelerated by the second half of the 1980s and international capital movements were liberalized with the decision of full convertibility in 1989. This situation triggered a rise in speculative capital flows into the national economy, thereby leading to an increase in public borrowing rates and the allocation of a significant part of public revenues to debt interest payments. To put it differently, the borrowing policy based on short-term capital inflows in the 1990-1994 period put the national economy under considerable pressure of domestic debt (Ümit, 2007: 116–120). Indeed, the public sector borrowing requirement to GDP ratio was 10.6 % in 1992 and rose to 12 % in 1993. The foreign trade deficit increased to 14.1 billion dollars by 72.7 % in 1993 and the export-import coverage ratio decreased to 52.1 % (Seyidoğlu, 2003: 146). Therefore, in 1993, Moody's downgraded Turkey's credit rating from Baa3 indicating an investment grade to Ba1 indicating a non-investment grade. This situation was one of the factors that triggered the 1994 crisis. S&P declared Turkey's credit rating to be BBB- on January 14, 1994, while Moody's declared it to be Ba1 on September 13, 1994. To put it differently, S&P and Moody's, one after the other, downgraded Turkey's credit rating. Fitch downgraded Turkey's credit rating from the investment grade BBB to the non-investment grade BB- in 1995 and to the speculative grade B+ in 1996. Turkey's credit score remained at the same level until 2004; S&P upgraded it to BB- in 2004, Moody's to Ba3, and Fitch to BB- in 2005.

In a nutshell, Turkey's low credit scores, especially in the 1990s, caused the country's borrowing in international markets rather than in markets that are more markets that are more convenient in terms of term structure and cost. This situation led to an increase in Turkey's domestic debts. The use of the national resources by the public sector failed to satisfy funding needs of the real sector of the economy, thereby causing the real sector to operate under high-interest rate conditions (Karagöl & Mıhçıokur, 2012: 18).

Additionally, Turkey witnessed a failed coup attempt on July 15, 2016. Thus, Moody's downgraded Turkey's credit rating from the investment grade Baa3 declared on May 16, 2013 to Ba1 by one notch on September 23, 2016. S&P also downgraded Turkey's credit rating from BB+ declared on February 7, 2014 to BB- by one notch on July 20, 2016. Fitch also downgraded Turkey's credit rating from BBB- with stable outlook declared on November 5, 2012 to BBB- with negative outlook on August 19, 2016 (BIANET, 2018).

4.5 Dot.com Crisis

The dot.com crisis, also referred to as the 2001 technology crisis in the United States (US), dragged the US economy into recession. The crisis broke out when a number of companies investing in internet technologies suddenly pulled out of the stock market and many technology companies went bankrupt (Kahraman, 2016: 16). To put it differently, the crisis was triggered when the technology-focused NASDAQ (National Association of Securities Dealers Automated Quotations) index peaked in value on March 10, 2000 but shortly after fell and the continuing decline caused a surge of panic (www.ekodiolog).

The AOL (American Online) internet media company is used to illustrate the dot.com crisis. The company value was $62 million in March 1992 when it was opened to the public and sharply rose to $19 billion in June 1998. The company value was $105 billion in July 1999 and rose to $161 billion in January 2000. Moody's rated AOL as B2 (highly speculative) in 1998 and upgraded its credit rating to the investment grade Baa1 before the outbreak of the dot.com crisis. The Moody's pre-crisis rating of the company as investable gave rise to criticisms of whether credit ratings given by rating agencies are reliable (Toraman & Yürük, 2014: 148).

4.6 Enron Scandal

The Enron Corporation, one of the world's largest natural gas pipeline companies, was ranked seventh by the Fortune magazine among the largest US companies shortly before its sudden bankruptcy on December 2, 2001 (Dinç & Cengiz,

2014: 229). Enron increased its assets from $10 billion to $63.4 billion within the 16-year period from its establishment in 1985 to its bankruptcy (Sağlar & Kandemir, 2007: 3). The bankruptcy of the company was recorded as the biggest corporate bankruptcies in the USA. The company went bankrupt because Arthur Andersen, the world's fifth largest accounting firm, hid the company's bad financial situation for three years (Dinç & Cengiz, 2014: 225). Additionally, Moody's, S&P, and Fitch rated the company at the investment-grade level four days before its bankruptcy. However, following the bankruptcy of the company, CRAs downgraded the company's credit rating. This situation led to the questioning of the credibility of CRAs (Çevik Tekin, 2016: 195).

4.7 Parmalat Scandal

Parmalat was established in 1961 in Parma, Italy, to operate in the sector of milk and dairy products. Having grown dramatically in the 1990s, Parmalat announced its sales revenue as €7.6 billion ($9 billion) by the end of 2002. In November 2000, S&P started to rate Parmalat as the fastest growing company and the eighth largest company in Italy. As of that time, S&P rated the company's credit score as the lowest investment grade BBB- and rated its ability to repay short-term debt as the investment grade AAA until December 2003 (Küçüksözen, 2004: 369–371). Due to irregularities in its accounting records, the company declared bankruptcy at the end of December 2003 and left a mark as one of the greatest scandals in Italy's history.

Parmalat underwent an independent audit in 2002 and disclosed the 2002 interim financial statements to the public. In December 2002, Merrill Lynch equity analysts examined the company's debt structure, financial performance, and disclosure of bond issues and downgraded their recommendation on Parmalat from "buy" to "sell". In March 2003, stock analysts (Lehman Brothers and other investment bank analysts) issued negative reports about the company. During the August-November 2003 period, the company offered $100 million of unsecured Senior Guaranteed Notes to American investors and falsely stated that it used its surplus cash of €2.9 billion to repurchase its outstanding debt securities. However, the company did not repurchase those debt securities. Thus, on September 15, 2003, S&P changed its credit rating for Parmalat from BBB- with positive outlook to stable outlook. On October 31, 2003, Parmalat's independent auditor published the company's audit report and declared that they lacked sufficient data to confirm the company's 6-month financial statements. On November 10, 2003, the company announced that it had an investment of $545 million on the Cayman Islands at the request of CONSOB (Commissione Nazionaleper le

Società e la Borsa) regulating the Italian securities market. The company also stated that it turned into cash other assets of billions of dollars with this investment as soon as possible to avoid any financial problems. On November 11, 2003, S&P downgraded its credit rating for Parmalat from BBB- with stable outlook to negative outlook. On December 9, 2003, the company defaulted on its bond debt of €150 million, thereby leading to a significant fall in the company's share prices. The company paid its bond debt the following week. Additionally, it became clear that the company did not actually pay its bond debt of €2.9 billion and did not have assets as opposed to what was reported in its financial statements. Therefore, S&P downgraded its credit rating to B+. On December 10, 2003, S&P downgraded its credit rating to C following the announcement of the company's actual financial position (Küçüksözen, 2004: 373–376).

4.8 United States Subprime Mortgage Crisis

The mortgage market consists of a primary market where mortgage loans are organized and a secondary market where mortgage securities and bonds are bought and sold. In the mortgage market, the process begins when an investor applies for a fixed or variable rate loan in a financial institution to buy a house, and an expert appraises the true value of the property and assesses the loan applicant's ability to repay the loan. Mortgage loan contracts made at the end of this process form the basis of the mortgage system. Problems that may arise in the course of the preparation of those contracts cause failures in the functioning of the system (BDDK, 2008: 13). The world's largest mortgage market is the US mortgage market with a size of $10 trillion. However, the collapse that started in the US housing market in 2007 caused a great instability in the financial markets. This instability turned into a liquidity crisis, spread throughout first developed and then developing countries, and eventually turned into a global financial crisis in 2008 (Çevik, 2016: 197). The cause of the US mortgage crisis was the boom in defaults and judicial sales with increasing interest rates due to the subprime mortgage lending issued regardless of borrowers' credit history with increased risk appetite when interest rates were low. This incident spread throughout the financial system through complex derivative instruments, thereby leading to fluctuations in the financial system. This process turned into a crisis after the second quarter of 2007; thus, defaults on mortgage loans negatively affected the entire system (BDDK, 2008: 45).

Many economists and politicians criticized CRAs' credit ratings during the 2007 crisis in the USA. Their criticism was that CRAs contributed to the credit market bubble in the 2007 crisis by giving high ratings to low-quality assets based on risky mortgage loans. In fact, CRAs have the greatest responsibility for pricing residential mortgage-backed securities (RMBS) in the mortgage markets. Credit ratings given by CRAs form the basis for pricing both in the credit market and in the securities market (Toraman & Yürük, 2014: 145). Thus, CRAs' credit ratings of 96 % of subprime mortgage bonds with A grades indicate that the foundation of the system was initially problematic. In the aftermath of the crisis, CRAs downgraded their credit ratings for mortgage bonds. Additionally, the increase in the number of investors seeking to withdraw money from funds made it difficult to turn those assets into cash. When CRAs' downgraded their inflated credit ratings of mortgage bonds, it raised questions about the system. Thus, the prices of investment instruments declined and billions of dollars of funds decreased in value (BDDK, 2008: 56-57). CRAs failed to perceive the magnitude of the problem in the fluctuation that began with losses on subprime mortgage securities. The US President's Working Group on Financial Markets declared that CRAs ignored the credit risk before the fluctuations that started in June 2007. CRAs inflated their credit ratings for high-cost and problem asset-backed securities. Thus, it was widely accepted by the public and experts that CRAs were at the center of the biggest financial crisis in the US history (Şimşek, 2009).

4.9 Bankruptcy of Lehman Brothers

New York-based Lehman Brothers was a long-established and large investment bank founded in the 19th century. The term 'too big to fail' was used to describe how large and strong Lehman Brothers was (Kılıçaslan & Giter, 2016: 76). However, the crisis, which started in 2007 with the collapse of housing markets and defaults on mortgage loans in the US, led to the bankruptcy of Lehman Brothers Holdings Inc. on September 15, 2008, marking the largest bankruptcy in history. The bankruptcy of Lehman Brothers was called the worst crisis since the Great Depression of 1929 and described as a 'once-in-a-century' crisis by Alan Greenspan, the then Chairman of the Federal Reserve of the United States.

The bankruptcy of Lehman Brothers triggered the 2008 global crisis and led to the questioning of the credibility of CRAs that rated Lehman Brothers with the highest investment grade AAA just one day before its bankruptcy. As in the case of the Enron scandal, many were blamed for the bankruptcy of Lehman Brothers. However, CRAS were considered responsible for the global crisis because they failed to properly rate a bank as large as to cause a global crisis (Kılıçaslan & Giter, 2016: 77).

In a nutshell, following the bankruptcy of Lehman Brothers, there has been an increase in criticisms leveled at CRAs as their early warning system failed and countries' changing credit ratings fell short of the market indicators. Although there was no significant change in the financial structure and economic indicators of the countries, CRAs announced multi-notch downgrades in their credit ratings, thereby leading to growing concerns about the accuracy and reliability of credit ratings. Additionally, this situation gave rise to a view that CRAs do not satisfyingly disclose their methods to the public. The methodology reports published by CRAs were found to be insufficient in clarifying credit rating period processes. It was claimed that CRAs were among the main actors that caused the deepening of the global crisis. This claim is justified by the fact CRAs' reports do not clarify the process of rating, the weighting of factors affecting scores, the approach to cross-country comparisons, and the extent to which qualitative factors such as a country's political situation affect their ratings and the fact that CRAs give higher ratings in exchange for higher payments (Karagöl & Mıhçıokur, 2012: 9).

4.10 Greece Debt Crisis

Although the global crisis began in 2007, it can be said that the crisis has shown its real impact in European economies starting from 2009 because since 2009, the volume of international trade has narrowed in Europe, public financing has become difficult, and the Eurozone crisis has intensified (Flassbeck and Lapavitsas, 2015: 14). Although this process has stagnated the economies of many European countries, countries facing with the problem of high current account deficit, high public debt, and real estate loan bubbles have further adversely affected. In this context, in the post-global crisis process in Europe, Greece with high public borrowing, Ireland and Spain with real estate credit bubbles, and Portugal, Spain and Ireland with current account deficit problems have become badly affected countries in the mentioned period (Beck and Peydro, 2015: 61). The sub-prime mortgage crisis in the United States has increased the risks on global loans, raised funding costs, and made countries with high public debt, such as Greece, difficult to outsource (Türk and Eraslan, 2016: 285–286). On the other hand, Greece was a country with a budget deficit of around 5 % of the average GDP in the period of 2001–2008, and could finance these deficits easily with foreign credits and foreign capital investments. However, in October 2009, the ratio of the budget deficit to GDP was announced to be 6 % in Greece (GUARDIAN, 2018), and the corresponding rate was revised to 12.7 % in November 2009. Since then, the European monetary union has faced a challenging situation that it has never encountered before, and public debt rates have increased very rapidly for many Eurozone countries (Santis, 2012: 2). Immediately after the revision of the budget deficit rate, government bond interest rates in Greece have started to increase and interest rates in the country have risen even more sharply with the CRA's downgrading of

Greek securities, and thus, a debt crisis in Greece, which suffers from high public debt and budget deficits, has become inevitable (Nelson et.al., 2011: 4–5).

Even though Greece has not been able to draw a macroeconomically successful framework following its participation in the Eurozone, it appears that the main CRAs rated it with relatively high grades in the past. For example, S & P increased the credit rating of the country 4 levels, raised to A+ from BBB from December 1999 to November 2003, Moody's increased the country's rating 3 levels, raised to A1 from Baa between December 1999 and November 2002, and finally Fitch increased the rating 4 levels, raised to A+ from BBB between June 1997 and October 2003. The stated high ratings allowed Greece to borrow from the foreign market with favorable maturities and interest rates, significantly reducing the cost of borrowing. However, as mentioned in the previous paragraph, after the revision of the ratio of the budget deficit to GDP in November 2009, CRAs have dropped Greece's credit rating in a short period of time repeatedly, in order to compensate for the credit rating they could not correctly do in the past. This practice exacerbated the economic downturn in the country, increased borrowing costs, and deepened the debt crisis. In brief, CRAs could not foresee but exacerbated Greece's debt crisis. On the other hand, it should also be emphasized that the manipulation of statistics on public financing in the country has triggered the crisis, and has also caused the incorrect rating. Because, the information considered in the rating is the information obtained from the company and the country, reflecting the truth is necessary for accurate rating (Bayar, 2015: 51–53).

5 Conclusion

The credit ratings market, which has an oligopolistic structure, is currently dominated by three major companies. The ratings given by these three companies to countries, financial institutions and securities significantly guide global trade volume and fund flows, so these companies play an important role in global markets. However, crises and bankruptcies of companies in world's different regions have led to questioning how effective these institutions are. Certain organizations, which seem to never bankrupt and are rated by CRAs with high grades, went out of business. It is confirmed by the crises that the financial structure of countries whose economic structure looks strong and rated by CRAs with high grades is not as strong as it might seem. Asia, Russia, Turkey, Greece, 2008 global crises and bankrupties of famous companies like Enron, Parmalat, and Lehman Brothers are the main events that raise questions about reliability of ratings. It is revealed with this study that CRAs could not foresee, even exacerbate financial crises with the consecutive rating downscales. However, this assessment does not ignore the need for CRAs and underlines the need for a more effective and more transparent rating activity that will remove the asymmetric information problem. In this context, the ratings of CRAs should be based on more reliable financial reports and controlling the rating sector with a better mechanism to mislead the markets is a necessity. In addition, it is thought that the market dominance of the three major credit rating agencies that dominate the market must be broken. Accordingly, studies have been started to create alternatives for credit rating agencies in many countries, especially in Europe, worldwide. If the stated measures are taken, it will be possible to anticipate the crises and to make the financial system healthier.

References

Akbulak, Y. (2012). "Kredi Derecelendirmesi veya Rating: Kavram ve Ölçütler", *Mali Çözüm Dergisi*, (Mayıs-Haziran 2012): 171–184.

Afonso, A., Gomes, P. & Rother, P. (2007). "What Hides Behind Sovereign Debt Ratings?", *European Central Bank Working Paper Series*, No. 711: 1–67.

Bayar, Y. (2015). "Kredi Derecelendirme Kuruluşları ve Yunanistan Borç Krizi", *Uluslararası İktisadi ve İdari İncelemeler Dergisi*, 8 (15): 41–58.

BDDK (2008). "ABD Mortgage Krizi", *Çalışma Tebliği*, (3, Ağustos 2008): 1–118.

Beck, T. & Peydro, J.L. (2015). "Five Years of Crisis (Resolotion)- Some Lessons", *The Eurozone Crisis: A Consensus View of the Cause and a Few Possible Solutions*, (Ed.) Richard Baldwin ve Francesco Giovazzi, Co-founded by European Union, EU, CEPR Press, 61–69.

BIANET (2018). http://bianet.org/bianet/siyaset/183205-1992-den-gunumuze-turkiye-nin-kredi-derecelendirme-karnesi, (14.05.2018).

Çevik Tekin, İ. (2016). "Kredi Derecelendirme Kuruluşlarının Öngöremedikleri Krizler Ve İflaslar", *Sosyal Bilimler Meslek Yüksekokulu Dergisi*, 19 (41.Yıl Özel Sayısı): 181–205.

Destraz, S. & Lahaye, R. (2012). "*Are Credit Ratings Trustworthy? Empirical Study on the Dependence of Corporate Defaults to Market Risk Within Investment Grade and Speculative Grade Range*", http://dx.doi.org/10.2139/ssrn.2056849, 29.05.2018).

Dinç, Y. & Cengiz, S. (2014). "Muhasebe Denetiminde Hata ve Hilenin Denetçi Etiği Açısından İncelenmesi: Enron Skandalı Örneği", Çankırı Karatekin Üniversitesi Sosyal Bilimler Enstitüsü Dergisi 5(1): 221–236.

Elkhoury, M. (2008). "Credit Rating Agencies and Their Potential Impact on Developing Countries", *UNCTAD Discussion Paper*, No.186: 1–33.

Engin, B. (2007). "Gelişmiş ve Yükselen Piyasalarda 1990 Sonrası Görülen Finansal Krizler ve Dünya Ekonomisi Üzerindeki Etkileri", *Sosyal Bilimler Dergisi*, No.2: 35–60.

European Securities and Markets Authority (2016). "Competition and Choice in the Credit Rating Industry". https://www.esma.europa.eu/sites/default/files/library/2015-1879_esma_cra_market_share_calculation.pdf, (29.05.2018).

FITCH, (2018). https://www.fitchratings.com/site/about, (29/05/2018).

Flassbeck, H. & Costas L., (2015), *Against the Troika Crisis and Austerity in the Eurozone*, Verso Press, London&NewYork.

Guardian, (2018). https://www.theguardian.com/business/2010/may/05/greece-debt-crisis-timeline, (30/05/2018).

Hamzo, I. H. (2007). *Kredi Risk Yönetimi*, İstanbul Üniversitesi Sosyal Bilimler Enstitüsü İşletme Anabilim Dalı Yüksek Lisans Tezi, İstanbul.

IMF (International Monetary Fund) World Economic Outlook (WEO) (May 1998). *"Global Repercussions of the Crises in Emerging Markets and Other Conjunctural Issues Chapter II"*. https://www.imf.org/en/Publications/WEO/Issues/2016/12/31/International-Financial-Contagion, (10.05.2018).

Karagöl, E. & Mıhçıokur, Ü. (2012). *Kredi Derecelendirme Kuruluşları Alternatif Arayışlar*, SETA Rapor, No.7, Eylül.

Kılıç, B. (1989). *Derecelendirme (Rating) İşlemi, ABD'de Tahvil Değerlendirme Süreci*, Sermaye Piyasası Kurulu Araştırma Raporu.

Kılıçaslan, H. & Giter, S. M. (2016). "Kredi Derecelendirme ve Ortaya Çıkan Sorunlar", *Maliye Araştırmaları Dergisi*, 2 (1): 61–81.

Kräussl, R. (2000). *Sovereign Ratings and Their Impact on Recent Financial Crises*, CFS Working Paper Series, No. 2000/04, Center for Financial Studies, Frankfurt, Mainz.

Kräussl, R. (2003). *Do Credit Rating Agencies Add to the Dynamics of Emerging Market Crises?* Center For Financial Studies, J.W.Goethe University Publication, Frankfurt, Mainz.

Küçüksözen, C. (2004). *Finansal Bilgi Manipülasyonu: Nedenleri, Yöntemleri, Amaçları, Teknikleri, Sonuçları ve İMKB Şirketleri Üzerine Ampirik Bir Çalışma*, PhD Thesis, Ankara University: Ankara.

MOODY'S (2018). https://www.moodys.com/Pages/atc.aspx, (28/05/2018).

Nelson, R., Belkin, P. & Mix, D. (2011). "Greece's Debt Crisis: Overview, Policy Responses, and Implications", *Congressional Research Service*, 7-5700, CRS Report for Congress.

OECD, (2010). *Competition and Credit Rating Agencies*, DAF/COMP 2010(29), https://www.oecd.org/competition/sectors/46825342.pdf, (29.05.2018).

Oktar, S. &Yüksel, S. (2015). "1998 Yılında Rusya'da Yaşanan Bankacılık Krizi ve Öncü Göstergeleri", *Marmara Üniversitesi İ.İ.B. Dergisi*, 37 (2): 327-340.

Özdinç, Ö. (1999). *Derecelendirme Sürecinde Ekonometrik Bir Değerlendirme*, SPK Yayınları, Yayın No: 130, Ankara.

Pascalis, F. D. (2017). *Credit Ratings and Market Over-Reliance: An International Legal Analysis*, Brill Publishing: Netherlands.

Resmi Gazete, (2012). *Derecelendirme Kuruluşlarının Yetkilendirilmesine ve Faaliyetlerine İlişkin Esaslar Hakkında Yönetmelik*, Bankacılık Düzenleme ve Denetleme Kurumu Yönetmeliği, Sayı:28267, 17 Nisan 2012.

Sağlar, J. & Kandemir, C. (2007). "Enron Olayı: Muhasebe Hilesi mi, Sistem Hatası mı?", *Çukurova Üniversitesi İİBF Dergisi*, 11(1), 20-39.

Santis, R., (2012). "The Euro Area Sovereign Debt Crisis Safe Haven, Credit Rating Agencies and The Spread of the Fever From Greece, Ireland and Portugal", *European Central Bank Working Paper Series*, No 1419: 1-61.

Seyidoğlu, H. (2003). "Uluslararası Mali Krizler, IMF Politikaları, Az Gelişmiş Ülkeler, Türkiye ve Dönüşüm Ekonomileri". *Doğuş Üniversitesi Dergisi*, 4 (2): 141-156.

Şimşek, E. (2009) *Krizin Ortaya Çıkmasında Kredi Değerlendirme Kuruluşlarının Rolü*, http://ekonomi.haber7.com/ekonomi/haber/397120-global-krizin-suclusu-derecelendirme-kuruluslari-mi, (19.05.2018).

SPK, (2018). http://www.spk.gov.tr/Sayfa/Index/6/10/2, (21/05/2018).

S&P (2018). https://www.spratings.com/en_US/what-we-do, (28.05.2018)

Toraman, C. & Yürük, M. F. (2014). "Kredi Derecelendirme Kuruluşları ve Finansal Krizlere Etkileri", *Bitlis Eren Üniversitesi Sosyal Bilimler Enstitüsü Dergisi*, 3 (1): 127-154.

Turan, Z. (2011). "Dünya'daki ve Türkiye' deki Krizlerin Ortaya Çıkış Nedenleri ve Ekonomik Kalkınmaya Etkisi", *Niğde Üniversitesi İİBF Dergisi*, 4 (1): 56-80.

Türk, A. & Eraslan, C. (2016). "Yunanistan Borç Krizinin Nedenleri Ve Sonuçları: Yapay Sinir Ağları ile Bir Analiz", *Gaziantep University Journal of Social Sciences*, 15 (2): 281-302.

Ümit, A. Ö. (2007). *Türkiye'de Bütçe Açığı ile Cari İşlemler Arasındaki İlişkilerin Zaman Serisi Analizi*, PhD. Thesis, Anadolu University: Eskişehir.

Yavaş, H. (2007). *1980 Sonrası Gelişmekte Olan Ülkelerde Yaşanan Finansal Krizler, Finansal Kriz Modelleri ve Çözüm Önerileri*, Kadir Has Üniversitesi Sosyal Bilimler Enstitüsü, Yayımlanmamış Yüksek Lisans Tezi.

Muhammet Yunus Şişman

Volatility and Foreign Direct Investment in MENA Region: A Spatial Panel Approach

1 Introduction

Foreign Direct Investment (FDI) has been well studied in the applied literature. This study attempts to explore the link between economic risk factors and foreign direct investment decisions by accounting for cross-country dependency effects. We consider volatility in macroeconomic environment as one of the key indicators for the risk at the country level. In particular, the study incorporates income, inflation, and exchange rate volatilities to proxy the level of macroeconomic risks. We analyzed the relationship between FDI activities and volatility measures for Middle Eastern and North African (MENA) countries as the region has been persistently volatile and less attractive for foreign investors.

The existing literature provides an extensive research on the determinants of FDI with less attention on the role of macroeconomic volatility and spatial interaction among host countries. Among others, Goldberg and Klein (1997) and Chakrabati and Scholnick (2002) documented the significant relationship between exchange rate volatility and FDI. Kiyota and Urata (2004), Ruiz (2005), and Ogunleye (2009) suggested that volatile currencies discourage foreign investors. However, Cushman (1988), Markusen (1995), and Pain and Van Welsum (2003) claimed a positive relationship justified with the export substituting role of FDI. The studies provided evidence that volatile exchange rates increase the FDI levels as the multinationals prefer serving foreign markets by local production over exports to minimize currency risk between headquarters and foreign markets.

Chan and Gemayel (2003) examined the macroeconomic volatility and FDI in MENA region by using risk index of the International Country Risk Guide (ICRG). The authors found that volatility measures of the index have a more significant impact on FDI than the index itself, thus they referred a detailed analysis on each component of the risk factors for future studies. In a more recent study, Udoh and Egwaikhide (2008) investigated the impact of the volatility of macroeconomic indicators on foreign investment in Nigeria. Their research suggested that exchange rate and inflation volatilities adversely affect FDI activities in the country.

The literature so far did not consider the cross-country dependency in the analysis. Blonigen et al. (2007) and Baltagi et al. (2007) explored spatial interactions among host countries and proposed spatial econometric models to incorporate the spillover effect in the parameter estimation. Abate (2015) considered the spatial dependence in the analysis to identify the relationship between macroeconomic instability and economic growth. The author suggested that volatility, particularly volatility in innovations, has a significant negative direct and spillover impact on economic growth. Therefore, the adverse effect of volatile economic environment in a country negatively affects the economic growth in its neighbors.

The contribution of this paper is to investigate the role of instability of macroeconomic indicators on foreign investments by incorporating cross-country dependency in MENA countries. The study allows us to identify how the direct and spillover effects affects the foreign direct investment decision. Thus, this study provides a more comprehensive and reliable outlook on the link between FDI and macroeconomic risk factors in MENA region.

Findings suggest that spatial spillover effects in exchange rate volatility, purchasing power, and market size play important role on the FDI. In addition, exchange rate and inflation volatilities, market size, trade openness, and purchasing power have positive effects on FDI in the region implying that attracting FDI in MENA countries is closely related to integration to global markets and economic dynamism.

The paper continues with a description of variables and data sources used in the analysis. The following section outlines the empirical framework including a detailed discussion on spatial dependence models. After the estimation results are presented, the paper concludes with policy recommendations and a summary of findings.

2 Data Description

MENA region is somehow a flexible geographic area. The World Bank defines 21 countries in the region while UNICEF includes 20 countries. Despite territorially more extended versions of the MENA region, the analysis focused mainly on 20 countries[1] and Turkey. Due to the lack of appropriate data a group of countries (i.e., Libya, Palestine, Sudan, and Syria) are excluded from the sample. Thus, remaining countries are examined between 2003 and 2011. Data is mostly available

---

1   These countries are Algeria, Bahrain, Egypt, Iran, Iraq, Israel, Jordan, Kuwait, Lebanon, Libya, Morocco, Oman, Palestine, Qatar, Saudi Arabia, Sudan, Syria, Tunisia, United Arab Emirates, and Yemen.

starting from 2003 for the countries in question. The analysis contains the period since 2011 to avoid the impact of mass chaotic political events (Arab Spring movements) across the region.

The main source of data is The World Bank's World Development Indicators (WDI). Population, gross domestic product per capita (GDP-PPP), inflation, and exchange rate data comes from WDI. Food and Agricultural Organization of the United Nations (FAO) provide a more comprehensive FDI data; we used inward FDI stock data from FAO. For trade openness, the sum of goods trade (exports and imports) is calculated as a percentage of GDP. Due to the incomplete data on both goods exports and imports, we combined United Nations Comtrade and WDI data[2]. Data series in absolute value levels such as inward FDI stock in million dollars, population, GDP per capita, and GDP volatility are all transformed into natural logarithm. The dependent variable is the natural logarithm of inward FDI stock. Some studies use annual inward FDI inflows as a proxy for FDI (Kyereboah-Colman and Agyire-Tettey, 2008; Renani and Mirfatah, 2012). Since inward FDI in MENA takes negative values for several years and negative values cannot be log-transformed, we used inward stock variable instead of inward flows. In order to capture the effects of market size, purchasing power, and trade openness as in traditional FDI models, three control variables are employed in the model. The first control variable is the natural logarithm of the total population as a proxy for market size. Second variable is the natural logarithm of GDP per capita at purchasing power parity (constant 2011 international dollars) as a proxy for purchasing power. The last control variable is the sum of goods exports and imports of a country divided by its GDP.

Exchange rate, income, and inflation volatilities are calculated in order to analyze the effects of macroeconomic volatility on FDI. There are mainly two methods employed in the literature for calculating the volatility measures: 1) standard deviation method (e.g., De Menil, 1999; Osinobi and Amaghionyeodime, 2009; Renani and Mirfatah, 2012) and 2) GARCH method (e.g., Engle, 1982; Bollerslev, 1986). For computational simplicity we used averages of five-year standard deviations of official exchange rate, GDP per capita, and annual GDP deflator to proxy the exchange rate, income, and inflation volatilities, respectively. Descriptive statistics for variables used in the model are given in Tab. 1.

---

2   Exports and imports data for Iran and Iraq are also interpolated for a few missing years.

Tab. 1: Descriptive statistics.

	Obs	Mean	Standard Deviation	Minimum	Maximum
Vdx	153	112.56	474.89	0	3424.28
Vinf	153	7.41	6.07	0.18	35.54
Vgdp	153	8.62	1.03	6.22	10.89
lnPPPper	153	9.92	0.97	8.24	11.79
Lnpop	153	16.22	1.39	13.41	18.24
Openness	153	0.87	0.42	0.20	2.33

3 Specification of Spatial Weight Matrix and Estimation Method

Linear relationship between macroeconomic volatility, control variables, and FDI are given in equation (1) below.

$$\ln fdistock = f(vdx, v\inf, vgdp, \ln PPPper, openness, \ln pop) \qquad (1)$$

In equation (1) *ln fdistock* is the natural logarithm of inward FDI stock; *vdx*, *vinf*, and *vgdp* represent exchange rate, inflation, and natural logarithm of GDP volatilities, respectively. ln *PPPper* is the natural logarithm of GDP per capita (PPP), *openness* is the sum of goods exports and imports divided by the GDP. *ln pop* is the natural logarithm of the total population.

In order to extend our analysis to examine the spillover effects of exchange rate, income, and inflation volatility in neighboring countries on FDI inflows to the host country, we employed spatial estimation methods. Spatial econometrics is a branch of econometrics which incorporates neighboring interactions between geographical units. Thereby, as stated in Tobler's (1970: 236) first law of geography: "*everything is related to everything else, but near things are more related than distant things*"; connections with the neighbors can be integrated into the analysis.

The focal point in spatial econometrics is the spatial weight matrix, which shows the relationship between geographical units. Spatial weight matrix, W, is a square matrix with $n \times n$ dimension where n is the number of geographical units. Neighborhood in spatial weight matrices can be defined as *contiguity* or *distance*. While distance-based spatial weight matrices use distances between geographical units, contiguity-based spatial weight matrices use shared boundaries. Contiguity-based spatial weight matrices become problematic for islands. A solution to this problem is to use a distance-based spatial weight matrix which ensures that every geographical unit has at least one neighbor, even an island. Distance between capital cities, centroids, or other theoretically relevant set of coordinates can be used for constructing a distance-based spatial weight matrix.

One of the most contradictive issue in applied spatial analysis is the specification of spatial weight matrices. Anselin et al. (2008) assert that the specification of spatial weight matrix is often *ad hoc*, if there is no formal theoretical reason for the structure of spatial interactions. Bell and Bockstael (2000) explicitly argue that the estimates and inferences are sensitive to small changes in spatial weight matrix. Additionally, Stakhovych and Bijmolt (2008) show that the correctly specified weight matrix is important for parameter estimation. They propose to apply weight matrix selection procedure that is based on "goodness of fit" criteria which increases the probability of identifying its true specification and benefits the researcher in terms of the precision of the parameter estimates compared with choosing the weight matrix arbitrarily. Harris, Moffat, and Kravtsova (2011) criticize this approach because "goodness of fit" criteria would not find correctly specified W, but only find local maximum among the competing spatial models. Elhorst (2010) points out that the monte carlo simulations carried out by Stakhovych and Bijmolt (2008) partly refute Harris, Moffat, and Kravtsova's (2011) critique. Lesage and Pace (2009) offers another criterion namely Bayesian posterior model probability. But mathematics of this approach will deter potential users (Elhorst, 2010).

As shown in Lesage and Pace (2009), Elhorst (2014) and Lesage (2014), if a particular independent variable in a cross-sectional unit changes, it will not only effect the dependent variable in that unit, it also effects the dependent variables in other neighboring cross-sectional units. The first effect explained above is the direct effect and the latter called the indirect effect.

Another important task that is routinely undertaken by researchers is the model selection. There are two main approaches for this task. The first one is to start with a non-spatial model and then to test whether or not this non-spatial model needs to be extended with spatial interactions. This approach is known as specific-to-general approach. The opposite approach is to start with a general model specification containing a series of simpler models that should ideally represent all the alternative hypothesis and test general specification for the alternative simpler models (Elhorst, 2014: 7). This approach is called general-to-specific approach.

There are four main types of spatial models widely used in the literature. These are Spatial Lag Model (SAR), Spatial Error Model (SEM), Spatial Durbin Model (SDM), and Spatial Autocorrelation Model (SAC). SAR incorporates spatially lagged variable on the right hand side of the model specification. SEM includes spatiality in the error term of the model. SDM considers both a spatially lagged dependent and independent variable on the right hand side of the model. And finally SAC is basically a combination of the SAR and SEM models. Lesage and Pace (2009) and Lesage (2014) put special emphasis on SDM and Spatial Durbin Error Model (SDEM). They argue that the SDM and SDEM models should be the mainstay models of practitioners. They assert that SDM model provides a general starting point for discussion of spatial regression model estimation since this model subsumes the SEM and the SAR. One advantage of SDM is that it produces unbiased coefficient estimates even if the true data generating process is a SAR or SEM. A state-of-the-art application of spatial econometrics should consider the SDM, and interpret its direct and indirect effects (Elhorst, 2010: 26).

Following Lesage and Pace (2009) and Elhorst (2010; 2014), five different models are specified to estimate equation 1. These specifications are SAR, SEM, SDM, and SAC models including a non-spatial model (Tab. 2). Six spatial weight matrices are tested for the data. Three of them are inverse distance matrix, two nearest neighbors, and three nearest neighbors. The other three spatial weight matrices are row normalized versions of the first three matrices.

The most common problem that practitioners face in econometric analysis is the presence of non-spherical errors. If this problem is not properly addressed, it can generate inefficiency and biasedness in the estimation. While serial correlation has long been an issue in panel data analysis, it can be said that cross-sectional dependency has recently gained attention. To address these problems, Parks-Kmenta suggested to use Feasible Generalized Least Squares (FGLS) estimator. However, this approach is appropriate if number of time periods, T, is greater than the number of cross sections, N (Reed and Ye, 2011; Hoechle, 2007). Beck-Katz proposed an alternative approach called Panel Corrected Standard Errors (PCSE) that performs better than FGLS (Reed and Ye, 2011). The weakness of PCSE is

Tab. 2: Variables and models.

y	ln *fdistock*
X	[$vdx, vinf, vgdp, \ln PPPper, \ln pop, openness$]
NSM	$y = \beta_0 + \beta_1 X + \varepsilon_{it}$
SAR	$y = \beta_0 + \rho W y + \beta_1 X + \varepsilon_{it}$
SEM	$y = \beta_0 + \beta_1 X + u_{it}; \quad u_{it} = \lambda W u_{it} + \varepsilon_{it}$
SDM	$y = \beta_0 + \rho W y + \beta_1 X + \theta W X + \varepsilon_{it}$
SAC	$y = \beta_0 + \rho W y + \beta_1 X + u_{it}; \quad u_{it} = \lambda W u_{it} + \varepsilon_{it}$

that when *N* gets bigger it underestimates the standard errors. Another approach to mitigate these problems is developed by Driscoll and Kraay (1998) which also performs better than PCSE if cross-sectional dimension, *N*, is large compared to time dimension, *T* (Hoechle, 2007). Driscoll-Kraay estimator produces robust standard error estimates where disturbances are being heteroscedastic, autocorrelated, and cross-sectionally dependent (Hoechle, 2007). Based on the diagnostic tests carried out in the study, Driscoll-Kraay robust standard errors are estimated for all models.

4 Estimation Results

In a linear model, we first need to check if a relationship exists among the explanatory variables. If they are correlated too much, that would be a sign of multicollinearity. Looking at correlations among the pairs of predictors to detect multicollinearity is a limited approach in that there is no bright line between "not too much correlated" or "too much correlated". Therefore, many practitioners rely on Variance Inflation Factor (VIF) and tolerance to detect multicollinearity. Tolerance and VIF values for our explanatory variables are given in Tab. 3.

As O'Brian (2007) noted, many scholarly articles and statistics textbooks commonly adopted the rule of thumb of 10 for VIF and 0.10 for Tolerance. When VIF is equal to or greater than 10, or Tolerance is equal to or lower than 0.10 there is said to be collinearity problem (O'Brian, 2007; Gujarati, 2004). In this respect, we can conclude collinearity is not an issue.

The next step in the analysis is to check for heteroscedasticity, serial correlation, and cross-sectional correlations. We conducted modified Wald test for group-wise heteroscedasticity, Bhargava Franzini and Narendranathan Durbin-Watson autocorrelation tests for the fixed effect models, and Pesaran CD test for cross-sectional dependency.

Tab. 3: VIF and tolerance.

Variable	VIF	Tolerance
Vdx	1.15	0.8731
Vinf	1.39	0.7188
Vgdp	1.36	0.7358
lnPPPper	2.08	0.4799
Lnpop	2.49	0.4014
Openness	1.39	0.7218
Mean VIF	1.64	

Tab. 4: Diagnostics tests.

	Test statistics
Modified Wald test for groupwise heteroscedasticity	5339.88*** (0.00)
Bhargava Franzini and Narendranathan Durbin-Watson autocorrelation test for fixed effect models	0.7194
Pesaran (2004) CD	8.68***(0.00)

dPL=1.83, dPU=1.88. H=50, T=10 and n=5, *** shows significance at 0.01.

Test results in Tab. 4 indicate that the data suffers from heteroscedasticity, serial correlation, and cross-sectional dependency. Therefore, Driscoll-Kraay robust standard errors are estimated to derive efficiency and unbiasedness in our estimation. There are four spatial estimation models and six spatial weight matrices competing to identify the best suited spatial model and weight matrix. Akaike Information Criteria (AIC), Schwarz Information Criteria (SIC), and Wald test statistics for different specifications are given in Tab. 5.

In Tab. 5, *Windis* is the inverse distance matrix, *Wnormdis* is the row normalized inverse distance matrix, *Wk2* and *Wk3* are the spatial weight matrices based on 2 and 3 nearest neighbors. *Wk2norm* and *Wk3norm* are the row normalized versions of *Wk2* and *Wk3*, respectively. It can be seen from the table above that the SDM model has the lowest AIC and SIC for most of the spatial weight matrices. SIC statistics for *Wnormdis, Wk2, Wk2norm,* and *Wk3norm* are lowest for SEM model. Additionally, we performed Wald test to compare SDM Model with SAR and SEM. According to Tab. 5, all Wald test statistics show the SDM model as the best fitting model for all spatial weight matrix specifications. For the spatial weight matrix, *Wk3* has the smallest AIC and SIC values. Thus, SDM model and *Wk3* matrix are preferred over the competing models and weight matrices.

In Tab. 6, non-spatial model shows positive and significant coefficients for ln*PPPper*, ln*pop*, and *openness*. These findings reveal that the purchasing power, market size, and international trade openness have positive effects on FDI in MENA which is consistent with the existing literature. Non-spatial model also reveals that the inflation volatility has a positive impact on FDI activities, while income volatility has a negative and statistically significant effect on FDI in MENA. Findings indicate that exchange rate volatility has an insignificant impact on foreign investments.

The spatial rho in the SDM model is significant at 1 percent level suggesting spatial dependency matters in FDI in the region. The indirect effects in SDM model (Tab. 6) show variables having spatial spillover effects on FDI. It turns out

Tab. 5: Goodness of fit statistics for alternative specifications.

		SAR	SEM	SDM	SAC
Windis	AIC	145.13	126.24	68.87	131.27
	SIC	223.92	150.48	147.66	210.06
	Wald SDM	467.70***	2483.46***		
Wnormdis	AIC	141.04	122.66	76.96	131.82
	SIC	219.83	146.90	155.76	210.61
	Wald SDM	379.14***	281.26***		
Wk2	AIC	151.85	125.70	74.02	124.13
	SIC	230.64	149.94	152.81	202.92
	Wald SDM	6134.92***	13601.95***		
Wk2norm	AIC	149.11	125.97	110.26	137.27
	SIC	227.90	150.21	189.05	216.07
	Wald SDM	681.08***	1026.63***		
Wk3	AIC	152.41	125.72	60.85	130.21
	SIC	231.20	149.97	139.64	209.00
	Wald SDM	445.84***	56275.99***		
Wk3norm	AIC	150.69	126.14	84.47	135.25
	SIC	229.48	150.38	163.26	214.05
	Wald SDM	593.30***	553.39***		

*** shows significance at 0.01.

that income and population variables have a positive spatial spillover effect on FDI. On the contrary, the exchange rate volatility has a negative spatial spillover effects on FDI.

Interpreting estimated coefficient is quite different in SDM model compared to standard fixed effects estimates. None of the direct, indirect, or total effects exactly coincide with standard β estimates of fixed effect model, which show the partial derivatives of dependent variable with respect to independent variables. Direct effects given in Tab. 6 can be interpreted as the effect of a change in an explanatory variable in a country on the dependent variable for the same country. In this respect, results suggest that the exchange rate volatility in a country positively effects FDI in that country. Same inferences can be made for inflation volatility, purchasing power, population, and trade openness.

In contrast to the direct effects, indirect effects or the spillover effects show the effect of a change of an explanatory variable in a country on the dependent variable in other countries. Tab. 6 indicates that the exchange rate volatility has

Tab. 6: Parameter estimates for non-spatial and spatial models.

	N: 153	Non-Spatial Fixed Effect (Driscoll-Kraay Std. Er.)		Spatial Durbin Model (Driscoll-Kraay Std. Er.)	
		Coefficient	Standard Er.	Coefficient	Standard Er.
Main	vdx	-0.000	(0.00)	0.000*	(0.00)
	vgdp	-0.210**	(0.08)	-0.090	(0.06)
	vinf	0.012**	(0.01)	0.017***	(0.00)
	lnPPPper	2.107***	(0.18)	1.244***	(0.13)
	lnpop	2.369***	(0.37)	1.745***	(0.10)
	openness	1.030***	(0.28)	0.574***	(0.21)
	constant	-48.780***	(6.69)		
Wx	vdx			-0.292***	(0.07)
	vgdp			-20.600	(23.14)
	vinf			0.066	(1.19)
	lnPPPper			272.253***	(102.31)
	lnpop			312.243***	(44.37)
	openness			-14.734	(28.48)
Spatial	rho			-34.927***	(9.80)
Direct	vdx			0.000***	(0.00)
	vgdp			-0.079	(0.06)
	vinf			0.017***	(0.00)
	lnPPPper			1.092***	(0.15)
	lnpop			1.563***	(0.06)
	openness			0.617***	(0.19)
Indirect	vdx			-0.002***	(0.00)
	vgdp			-0.084	(0.13)
	vinf			-0.003	(0.01)
	lnPPPper			1.202**	(0.58)
	lnpop			1.373***	(0.25)
	openness			-0.228	(0.16)
Total	vdx			-0.001***	(0.00)
	vgdp			-0.162	(0.14)
	vinf			0.015***	(0.01)
	lnPPPper			2.294***	(0.49)
	lnpop			2.937***	(0.21)
	openness			0.389*	(0.23)
	R-sqr	0.72		0.86	

*p<0.10 **p<0.05 ***p<0.01

negative spillover effects on FDI in MENA. This finding suggests that when the exchange rate becomes more volatile in neighboring countries, FDI in host country decreases. Income volatility has an insignificant indirect effect. In contrast to direct effects, inflation volatility has an insignificant coefficient and shows no spillover effect. Market size and purchasing power of neighboring countries have a positive indirect effect on FDI suggesting an increase in the market size of a country stimulates FDI activities in its neighboring countries. Trade openness has a positive direct effect, but does not show spillover effects on FDI in MENA. The total effect of trade openness on FDI is again positive which shows that the direct effect of the trade openness dominates its indirect effect.

In terms of the total effects, exchange rate volatility is insignificant for nonspatial panel estimation but has a negative and significant effect on FDI in SDM estimation. On the contrary, income volatility has a significant impact on FDI in the non-spatial model estimation but it has insignificant direct, indirect, and total effects in SDM. Also market size, purchasing power, and trade openness have positive and significant effects on FDI for both non-spatial and SDM models. SDM model may reveal a more comprehensive outlook on the link between FDI and volatility as it allows us to identify the individual direct, indirect, and total effects.

5 Conclusions and Recommendations

This research aims to investigate the link between macroeconomic risk factors such as income, inflation, and exchange rate volatility and FDI by accounting spatial spillover effects in MENA region. The emphasis is especially given on the spatial interaction effects as foreign investors consider cross-country interactions, and the choice of FDI location is influenced by location-specific characteristics. Findings indicate that cross-country dependence matters for FDI in the region. While non-spatial model shows insignificant effect, Spatial Durbin Model shows positive direct effect but negative indirect effect. A group of MENA countries (i.e., Bahrain, Jordan, Lebanon, Oman, Qatar, Saudi Arabia, and United Arab Emirates) have adopted the fixed exchange rate regime. Therefore, volatility in those countries are very limited. When countries increasingly involve in free market economy, there is a shift towards greater exchange rate flexibility. Thus, the countries adopted floating exchange rate regimes potentially experience higher exchange rate volatility in MENA and are more integrated to free market economy. In this respect, the positive sign of direct effect may imply that the foreign investors preferred to invest in more globally integrated economies. This implication can be used to explain the indirect effects. Results indicate a negative indirect effect of exchange rate volatility on FDI suggesting that country i has adopted fixed exchange rate and its neighbors have adopted floating exchange rate, since exchange rates are more volatile in a floating regime, countries with floating exchange rate – also globally more integrated – attract FDI more than country i. Finally, the total effect in exchange rate volatility is negative reflecting the indirect spillover effects as the dominant factor.

Findings reveal that inflation volatility shows positive direct effect on FDI in MENA region. The GDP deflator employed as a proxy for inflation has positive sign for most of the countries and periods in our sample. This reflects an upward trend in inflation on average. One of the main reasons of inflation is the increase in the consumption expenditures. Moderate increases in inflation signals an economic recovery or greater economic dynamism. This dynamism in the economy will have a positive effect and attract FDI. Another thing that needs consideration is that the income volatility is significant for non-spatial model but insignificant in SDM. The neighboring effect potentially causes a better estimation accuracy of spatial models compared to OLS models. While direct and indirect effects of purchasing power and market size are positive and significant, trade openness has positive direct effect.

In summary, findings suggest that internationally more integrated economies having a larger domestic market size and higher purchasing power attract more FDI activities in MENA countries. Policy makers should promote economic integration through more liberal trade policies and exchange rate regimes. It is also important to keep inflation rate at moderate levels, which will help to maintain economic growth and economic dynamism which in return attracts more FDI activities in the region. Future research may incorporate political and institutional variables in addition to economic variables to potentially yield a more comprehensive analysis on the relationship between macroeconomic risks and FDI in MENA region.

References

Abate, Girum D. 2016. "On the Link Between Volatility and Growth: A Spatial Econometrics Approach." *Spatial Economic Analysis* 11 (1): 27–45.

Anselin, L., Le Gallo J., & Hubert J. 2008. "Spatial Panel Econometrics." In *The Econometrics of Panel Data Fundamentals and Recent Developments in Theory and Practice*, edited by L. Matyas, and P. Sevestre. Springer-Verlag, Berlin, 625–660.

Baltagi, Badi H., Egger P., & Pfaffermayr, M. 2007. "Estimating Models of Complex FDI: Are There Third-Country Effects?" *Journal of Econometrics* 140: 260–281.

Bell, K. P., & Bockstael. N.E. 2000. "Applying the Generalized-Moments Estimation Approach to Spatial Problems Involving Micro-Level Data." *Review of Economics and Statistics* 82 (1): 72–82.

Blonigen, Bruce A., Ronald B. Davies, Glen R. Waddell, and Helen T. Naughton. 2007. "FDI in Space: Spatial Autoregressive Relationships in Foreign Direct Investment." *European Economic Review* 51: 1303–1325.

Bollerslev, Tim. 1986. "Generalized Autoregressive Conditional Heteroskedasticity." *Journal of Econometrics* 31: 307–27.

Chakrabarti, Rajesh, and Barry Scholnick. 2002. "Exchange-Rate Expectations and Foreign Direct Investment Flows." *Weltwirtschaftliches Archiv* 138: 1-21.

Chan, Kitty K., and Edward R. Gemayel. 2003. Macroeconomic Instability and the Pattern of FDI in the MENA Region. Paper presented in 10th Economic Research Forum Conference, Morocco.

Cushman, David O. 1988. "Exchange-Rate Uncertainty and Foreign Direct Investment in the United States." *Weltwirtschaftliches Archiv* 124: 322–336.

De Menil, Georges. 1999. "Real Capital Market Integration in the EU: How Far Has it Gone? What Will the Effect of the Euro be?" *Economic Policy* 28:165–189.

Driscoll, John C., and Aart C. Kraay. 1998. "Consistent Covariance Matrix Estimation with Spatially Dependent Panel Data." *The Review of Economics and Statistics* 80 (4): 549–560.

Elhorst, J. Paul. 2010. "Applied Spatial Econometrics: Raising the Bar." *Spatial Economic Analysis* 5 (1): 9–28.

Elhorst, J. Paul. 2014. *Spatial Econometrics from Cross-Sectional Data to Spatial Panels.* Springer, Groningen, The Netherlands.

Engel, Robert F. 1982. "Autoregressive Conditional Heteroscedasticity with Estimates of the Variance of the United Kingdom Inflation." *Econometrica* 50: 987–1007.

Food and Agricultural Organization of the United Nations. FAOSTAT. Accessed 04 December 2016. http://www.fao.org/faostat/en/#data/FDI

Goldberg, Linda S., and Michael W. Klein. 1997. *Foreign Direct Investment, Trade and Real Exchange Rate Linkages in Southeast Asia and Latin America.* Paper No. 6344. National Bureau of Economic Research.

Gujarati, Damodar. 2004. *Basic Econometrics*, 4. Ed. The McGraw-Hill Companies. New York.

Harris, Richard, John Moffat, and Victoria Kravtsova. 2011. "In search of 'W'" *Spatial Economic Analysis* 6 (3): 249–270.

Hoechle, Daniel. 2007. "Robust Standard Errors for Panel Regressions with Cross-Sectional Dependence." *The Stata Journal* 7 (3): 281–312.

Kiyota, Kozo, and Shujiro Urata. 2004. "Exchange Rate, Exchange Rate Volatility and Foreign Direct Investment." *The World Economy* 27 (10): 1501–1536.

Kyereboah-Coleman, Anthony, and Kwame F. Agyire-Tettey. 2008. "Effect of Exchange Rate Volatility on Foreign Direct Investment in Sub-Saharan Africa The Case of Ghana." *The Journal of Risk Finance* 9 (1): 52–70.

Lesage, James P. 2014. "What Regional Scientists Need to Know About Spatial Econometrics." Accessed 10 February 2017. https://papers.ssrn.com/sol3/papers.cfm?abstract_id= 2420725.

Lesage, James, and R. Kelley Pace. 2009. *Introduction to Spatial Econometrics*, CRC Press Taylor & Francis Group. Boca Raton.

Markusen, James R. 1995. "The Boundaries of Multinational Enterprises and the Theory of International Trade." *Journal of Economic Review* 9: 169–189.

O'Brian, Robert M. 2007. "A Caution Regarding Rules of Thumb for Variance Inflation Factors." *Quality & Quantity* 41: 673–690.

Ogunleye, Eric Kehinde. 2009. "Exchange Rate Volatility and Foreign Direct Investment in Sub-Saharan Africa: Evidence from Nigeria and South Africa." Accessed 25 January 2017. http://www.csae.ox.ac.uk/conferences/2009-EdiA/papers/196-Ogunleye.pdf

Osinubi, Tokunbo S., and Lloyd A. Amaghionyeodiwe. 2009. "Foreign Direct Investment and Exchange Rate Volatility in Nigeria." *International Journal of Applied Econometrics and Quantitative Studies* 6 (2): 83–116.

Pain, Nigel, and Desiree Van Welsum. 2003. "Untying the Gordian Knot: The Multiple Links between Exchange Rates and Foreign Direct Investment." *Journal of Common Market Studies* 41: 823–846.

Reed, W. Robert, and Haichun Ye. 2011. "Which Panel Data Estimator Should I Use?" *Applied Economics* 43: 985–1000.

Sharifi-Renani, Hosein, and Maryam Mirfatah. 2012. "The Impact of Exchange Rate Volatility on Foreign Direct Investment in Iran." *Procedia Economics and Finance* 1: 365–373.

Stakhovych, Stanislav, and Tammo H. A. Bijmolt. 2008. "Specification of Spatial Models: A Simulation Study on Weight Matrices." *Papers in Regional Science* 88 (2): 389–408.

World Development Indicators Databank. Accessed 14 December 2016. http://databank.worldbank.org/data/reports.aspx?source=world-development-indicators.

Tobler, Waldo. 1970. "A Computer Movie Simulating Urban Growth in the Detroit Region." *Economic Geography* 46: 234–240.

Udoh, Elijah, and Festus o. Egwaikhide. 2008. "Exchange Rate Volatility, Inflation Uncertainty and Foreign Direct Investment in Nigeria." *Botswana Journal of Economics* 5 (7): 14–31.

United Nations. Comtrade Database. Accessed 11 December 2016. https://comtrade.un.org/data

Bilge Nur Öztürk and Tolga Öztürk

An Alternative Lifestyle Practice in a Globalizing World: Voluntary Simplicity and Cittaslow

1 Introduction

From the end of the 19th century, after the Industrial Revolution, capitalism first began to expropriate people. This expropriation process coincided with the birth of modern capitalist cities. Modern man has had to face the concept of individualization by encouraging modern cities. Individualization is expressed as the synonym of liberation, and this concept has romanticized by capitalism. But the economic system, in the name of emancipation to humans, has in fact led to more loneliness, more consumption, and a senseless quick experience of the people of the modern city torn from the earth. In this context, the modern life in question has offered people selfishness focused on consumption. People in modern urban life are becoming more isolated and unhappy.

Modern man has begun to realize that the sources of unhappiness originate in modern urban life. At least we can say that this awareness occurs within a certain group. In this context, contrary to the rapid life of modern urban life, currents advocating a slower life have begun to emerge. The first organized movement that emerged in Italy defended the necessity of differentiating the urban life primarily by recognizing and repairing the destructive effects of the capitalist process on people. In order to reverse the harms of modern cities to human beings, the slow cities tend to encourage the slow movement and sharing, genuine human relationships, slow food rather than selfishness, insincere relationships, and fast food. In this context, the Cittaslow cities appeared like an oasis in the gigantic deserts of modern cities. It is one of the most concrete movements since the 19th century to overcome the deadlocks of modern man and modern cities. In this context, the Cittaslow Movement is aiming a more economical and sustainable world, and offers as well a more peaceful life for the people.

2 Sustainable Consumption

Five percent of the world consume 25 % of all world resources and pollute the same rate. Every day more than one million tons of toxic waste is thrown into the environment. Thirty percent of natural resources are unfortunately lost

(Odabaşı, 2013). The current consumption of mankind now requires 1.4 of the world's resources. If people all over the world have a consumption at the level of the UK, there are 3.4 more world-like planets; if there is a consumption at the level of the United States, there are five more world-like planets needed (Global Footprint Network, 2009).

The act of consumption is itself associated with happiness. Contrary to what you might think, the pleasure of consuming and buying is temporary and has no lasting effects. According to research, individuals with severe desire for income and prosperity are less happy, have lower self-esteem, have higher levels of anxiety, and have weaker social relationships (Quelch & Jocz, 2007: 49–60). Psychologists have found that consumption-dominant lifestyles are detrimental to happiness and enjoyment in life (Myers, 2000: 2419). Even economically, it creates financial stress as it will cause excessive spending (Schor, 1999: 19).

Especially in some countries, we can say that people consume far more than real needs. Consumer type underlying this disproportionate consumption can be considered selfish and not interested in over-consumption results. As a result of this consumption mode, the world is gradually turning into a waste repository, climates are changing, some living species are extinct, air-water-soil quality is gradually decreasing, in other words, our vital future is being dragged towards a critical point.

The adverse effects of rapid and excessive consumption behaviors, both in the individuals and in the community, have led to a general awareness to this issue. Some concepts that are shaped by the desire to reverse this situation, or at least to stop it, have begun to enter our lives. Thus, new concepts such as "green growth", "environment friendly companies", "green marketing", "green consumer", and "ethical consumption" have emerged. In fact, these concepts have been developed to reach a sustainable world goal from the broadest perspective. First of all, for a sustainable world, it is necessary to control excessive and rapid consumption actions. A transformation towards "sustainable consumption" action should take place in individuals and communities.

Sustainable consumptions can be expressed, using our products and services to increase our quality of life, not to risk the needs of future generations and to be responsive to their needs (Odabaşı, 2013: 186). Sustainable consumption is a multidimensional concept. In general, it can be thought of as post-consumer behavior, energy-efficient behavior, and ecologically-informed vendor behavior (Leary vd., 2014: 1954).

Excessive consumption is a non-productive and unsustainable habit. For example, the United States makes 32 % of total global spending in 2006, with a population corresponding to 5 % of the world's population (Assadourian, 2010). Unlimited consumption is not possible in a world where there are ecological limits. As soon as they realize this, businesses have begun to include the green paradigm in their strategies and production processes (Strizhakova & Coulter, 2013). In order to imagine the idea of sustainable consumption, consumers need to be more attentive to consumption actions and producers must be rigorous in the same direction. In addition, advertising and communication efforts have an important place in creating sustainable lifestyles, avoiding excessive consumption (Sheth vd., 2011: 33). The outcome of environmental sustainability is simple: If everyone realizes that the ecosystem and natural resources are limited, economic decisions will be shaped in this direction and economic actions will be environmentally sustainable. For this, it is necessary to focus on values such as less use of resources, less waste disposal, and less pollution (Sharma vd., 2010).

2.1 Green Consumer

From the idea of sustainable consumption action, a consumer form, defined as a green consumer, has begun to take shape. The green consumer can be considered as a consumer who is susceptible to environmental issues and concerned about environmental issues (Kilbourne vd., 2002: 194). According to another definition, a green consumer is defined as a consumer who voluntarily and actively plays a role in the protection of nature / environment and who is able to direct production and consumption and to assess whether after-consumption wastes harm the environment (Odabaşı, 2013). In this context, this direction to production should be considered in the sense that consumers play an active role in non-environmentally production processes. The turnaround process of consumption will be achieved through the preference of green products and the elimination of non-green ones. Today, 75 % of consumers describe themselves as green consumers and prefer environmentally friendly products (Saad, 2006). Concepts such as green consumer, environmentally conscious consumer, ethical consumer, and socially responsible consumer that emerged with the sustainability efforts can be separated from each other with a few nuances, while pointing to a similar consumer model.

The ultimate point that green consumers come to in terms of the developmental stage is the consumers who can change their lifestyles in this context. In other words, this change in consumption behavior, which began in green consumers, has led to the formation of new consumption patterns and consumer forms that devote all their lifestyles to less to consume and more to produce.

2.1.1 A Green Consumer Movement: Voluntary Simplicity

As discussions progressed, opinions were raised that human beings should limit their consumption. For example, environmental awareness has encouraged consumers to reduce energy consumption (Kasulis vd., 1981). Consumers are now seeking answers to new questions with new awareness (Kotler, 2011): *Are we eating too much food? Is it the wrong kind of food? Are we driving a fuel-efficient car? Do we even need to own a car? Can we save more energy in the running of our home by more efficient lighting, solar panels, and other steps? Can we sort trash more efficiently into paper, cans and bottles, and waste?* These questions have caused people to think about their own consumption practices. Lifestyle advocates for less consumption have begun to evolve. Moving from this, a life far from materialistic values is not a more unhappy experience. Personal happiness and quality of life are not increasingly associated with excessive consumption (Kotler, 2011).

According to a survey (Gerzema, 2010), 62 % of Americans agree that *"Since the recession, I realize I am happier with a simpler, more down to basics lifestyle".* Seventy-seven percent stated that their time was more important than how much they earned. Consumers who are defined as lifestyles of health and sustainability (LOHAS) and embracing sustainability as a lifestyle have begun to emerge. Organic foods, energy efficiency facilities, solar panels, alternative medicines, and ecotourism markets have been formed for these people (Kotler, 2011).

As long as production and consumption are in large masses and more than real needs, the aforementioned green sensitivity will not be enough to protect world resources. Consumed products, even if they are green, will be relatively harmless to nature in all cases where they are consumed more than needed because they cannot be completely innocent towards the environment. Defined for real green consumption, it is a simple life form that people and institutions do not consume more than their needs. This lifestyle emerges as a voluntary simplicity movement.

Most of the 20th century included consumer movements related to defining and reaching consumer rights. However, a wave of anti-consumerism has shown itself. There has been a variety of worries over the public, excessive consumption, abuse of developing nations, and intense advertising. These concerns have resulted in anti-consumerism, combined with ecological / ethical concerns, reducing consumption and boycotting certain product categories. National initiatives such as "the International Buy Nothing Day (IBND)" and "Turn Off TV Week" have occurred. All these are indicative of a new set of consumers created by people who voluntarily want to simplify (Craig-Lees & Hill, 2002).

These people, who are defined as volunteer simplifiers, are the individuals who choose to live and to confine less (Elgin & Mitchell, 1977). Volunteer simplifiers are shortly those who are satisfied with less consumption, who loves simpleness, who are looking for alternative lifestyles, seeking high personal health, education, and satisfaction. In addition, they can be thought of as a set of volunteers who are willing to reduce their expenditure, trying to increase their control over their daily activities, and begin to shape a new lifestyle that is shaped by values such as humanism, self-determination, environmentalism, spirituality, and personal development. Persons can adopt such a lifestyle because of their environmental concerns, their religion, personal health, and happiness criteria. These people are also community-focused. In addition, they see it as an exchange to increase their time and consume less. While they consume, they pay attention to buy cheaper products and to buy less (Craig-Lees & Hill, 2002).

Lifestyle research involving voluntary simplification is often found in social psychology studies. The voluntary simplification model is not to have less than everything. It means having less than the material things and having more than the non-material things (Shama, 1996). In fact, it can be thought of as an effort to bring the lives of people under their control.

Voluntary simplification can be conceptualized as a lifestyle that offers sustainable consumption practices. However, sustainable consumption is not the only issue in this trend. Resource conservation, waste reduction, ecological effects are also involved in this simplification effort. In general, it can be defined as a belief system that offers an alternative culture practice to a consumption-based lifestyle. It aims to achieve this by producing systems that attach importance to nature, human and personal development, not just by limiting consumption but by emphasizing satisfaction from the non-material things (McDonald vd., 2006: 518).

3 Slow Food

Consumers' desire to simplify, slow down and spend more time on themselves as mentioned above, has helped shape the "slow food" movement. Eating habits that are familiar to fast-paced practices have turned into a style in which people cannot come and chat, cannot socialize, and cannot enjoy the moment when they are eating. This change request for the current state of the world, organized in many countries, has led to the emergence of non-profit and non-governmental organization Slow Food Movement.

"Slow food" was initiated as an international project under the leadership of Carlo Petrini in 1986, in order to encourage more consumption of traditional, regional foods as a solution to fast food and its consumption practices. But it is more than just trying to change eating habits. It should be considered of our food as a lifestyle that we perceive as social, ethical, political, environmental, and spiritual. Spending time to cook their own meals, cultural and traditional dishes being able to transfer to future generations is to understand with social and cultural perspectives which is better than unhealthy produced fast food. In other words, the Slow Food Movement has emerged as a movement that has the principle of enjoying local flavors, respecting nature, eating comfortably and healthily, conscious eating, even it is possible to grow food for themselves, getting taste from food, socializing with food (Sırım, 2012).

The primary purpose of the Slow Food Movement is to preserve traditional cuisine, food culture, from globalization, fast food and fast life culture (Güven, 2011). At the same time, the Slow Food Movement advocates the protection of food and agriculture biodiversity around the world and the preservation of all animal breeds and vegetable species on Earth. The standardization of flavors (as in fast food culture) is a condition that should not exist in the Slow Food philosophy. It seeks to ensure the sustainability of cultural identities in which consumers become conscious of food and gastronomic traditions (Yurtseven vd., 2010).

The Slow Food Movement is aiming to bring the kinds of food that need to be craftsmanship, which are about to be forgotten, into the face of the day and taking them in the global market from economic point of view. It helps small producers find each other and overcome bureaucratic obstacles. The forerunner of the Slow Food Movement, Carlo Petrini, stated that this movement was not against globalization and emphasized that the advocated was actually 'virtuous globalization'. The fact that traditional flavors are durable and robust for going overseas, the use of advanced technology, and the increasing awareness of the product and the provision of global recognition are important elements of 'virtuous globalization' (Çakır & Çakır, 2015).

The Slow Food philosophy is based on the concept of region. With the words of Carlo Petrini, the founder of the Slow Food Movement, the concept of the region is based on the combination of the natural factors (soil, water, land, sea level altitude, vegetation cover, microclimate) and the cultivated, built, and cooked food, and in every agricultural locality is the human factor (product cultivation tradition and practices) (Sezgin & Ünüvar, 2011: 117).

Good food, which is one of the basic elements of the Slow Food movement, is a fresh, delicious, seasonal, and naturally pristine fruit which is the fruit of the sufficiency of the producer and its raw materials and production methods as a part of the local culturally appealing, unique flavor, odor, color, shape, and texture. Another element, *clean food*, is the preservation of the health of the consumer and the producer, where ecosystems and biodiversity are protected, including consumption, in the agro-industrial production chain, where sustainable agriculture, farming, processing, marketing, and consumption practices are seriously addressed. *Fair food* is the element that expresses food created by respect for humanity and its rights, created by respect for cultural diversity and tradition, with sympathy and solidarity practice, pursuing balanced global economies where social justice is provided by the ability to produce sufficient prizes (Slow Food, 2018a).

The activities that the Slow Food Movement has launched in Italy since 1986 are continuing today with the opening of Germany in 1992, Switzerland in 1993, the USA in 2000, Japan in 2004, the UK in 2006, and the Netherlands in 2008. Today, Slow Food represents a global movement involving thousands of projects and millions of people in more than 160 countries (Slow Food, 2018b). The Slow Food movement has also highly begun to see support and interest in Turkey.

4 Cittaslow

Here, the concept of "Cittaslow" has been shaped on the basis of all this slow life, slow food, protection of the urban spirit, and sustainable development activities. The Cittaslow movement occurred in 1999 (Cittaslow, 2018a). Today the Cittaslow movement, which spreads to 236 cities in 30 countries / regions, moves the Slow Food philosophy to the urban dimension (Cittaslow, 2018b).

> "With the impact of globalization, cities have become non-self sufficient living spaces that are fast-paced, fast-consuming places. Cities have evolved from places where people with established intentions live together in a safe environment, to spaces designed for people to move faster and work faster. The limited life, which is the most important value of human being is presented as indispensable of modern life, with unhealthy foods, air pollution, traffic, loneliness and consumption. Popular cultures also support life without the time to live, going to work quickly with a car, drinking coffee while walking because of having no time, instead of enjoying the food because it is something that needs to be catch up, it is "fed" quickly, do not know their neighbors or local trades, which expresses modern human being is clearly not sustainable" (Cittaslow, 2018a).

Moving from people's consumption-focused and fast-paced life, the increase in general unhappiness and health problems revealed the Cittaslow Movement with a search for a new life-saving. Cittaslow philosophy advocates that life be experienced at a rate that can be enjoyed from life. The Cittaslow cities have become a realistic alternative to the modern cities, where people can communicate with each other, can socialize, be self-sufficient, sustainable, possess the traditions and customs of crafts, nature, but also use renewable energy resources (Cittaslow, 2018a).

Cittaslow movement originates from the Slow Food philosophy, has started to see quite interest in Turkey. The new Cittaslows are formed with the acceptance of the Cittaslow by applying to the International Cittaslow Union, as the main criteria of the more sustainable, traditionally preserved areas of slower living are maintained. It is necessary that the population of the cities that will apply to the

Tab. 1: Citta Slow Cities in Europe. **Resource:** Zawadzka, 2017: 96.

Population	<1000	1001–2000	2001–3000	3001–4000	4001–5000	5001–10.000	10.001–15.000	15.001–20.000	20.001–25.000	25.001–30.000	30.001–35.000	35.001–40.000	40.001–45.000	45.001–50.000	>50.000	Total Cities
Italy	6	7	8	8	6	25	9	3	6	2	1		1		1	83
Poland			3	1	3	4	8	3	4	1						27
Germany		1		1	3	3	2	3	4							17
Turkey						3	2	2	1	1	3	1			1	14
Holland						3		4	1	1	1					10
Spain	1		2	1		3		1								8
France	2	2	1	1	1											7
Belgium					1	2	1		1	1						6
Portugal							1		1	2	1	1				6
Great Britain				1		1	2				1					5
Norway	1	1		1		1										4
Austria						2	1									3
Cyprus				1		1	1									3
Denmark									1	1						2
Finland						1										1
Hungary														1		1
Iceland	1															1
Ireland					1											1
Sweden							1									1
Switzerland								1								1
Total Cities	10	11	14	15	15	49	28	18	18	8	7	2	2	2	2	201

Union is less than 50,000 and that it is compatible with the Cittaslow philosophy of the city administration (Cittaslow, 2018c). Tab. 1 contains the Cittaslows in Europe. As can be seen from Tab. 1, Turkey with 14 Cittaslows is the 4th country among 20 countries in Europe.

4.1 Cittaslow Cities in Turkey

The 14 Cittaslow cities in Turkey are as follows: Akyaka (Muğla), Eğirdir (Isparta), Gökçeada (Çanakkale), Gerze (Sinop), Göynük (Bolu), Halfeti (Şanlıurfa), Perşembe (Ordu), Şavşat (Artvin), Seferihisar (İzmir), Taraklı (Sakarya), Uzundere (Erzurum), Vize (Kırklareli), Yalvaç (Isparta), and Yenipazar (Aydın). Each Cittaslow has its own unique texture and unique features. Below is information pertaining to the 14 Cittaslows in Turkey (Cittaslow, 2018d)[1]:

4.1.1 Akyaka (Muğla)

Akyaka's Vision can be expressed as "*Akyaka, which respects the nature, preserves its architectural structure, has a strong service infrastructure, sustainable and alternative tourism-oriented people can act together in unity and togetherness*".

Akyaka is one of Turkey's 14 Cittaslow cities which has silence and calmness, mountains covered with lush green forests, aquarium-like ponds, unique underwater flora in these ponds, clear blue sea, forest camp, abundant water, historical buildings, wood houses in its unique architectural feature, otters which are a part of natural life, eel fish, Mediterranean seals, flamingos, and storks.

4.1.2 Eğirdir (Isparta)

Eğirdir is located in Isparta province, which has unique features such as: Eğirdir Lake, which changes color every time of day and every season, Can Island where its deed given by the residents of Eğirdir to Atatürk, The Turkish Armed Forces' Mountain Commando School, Kasnak oak (Quercus Vulcanica) and shallow tree forest which are rarely encountered in the world, Turkey's leading Osteopathic Hospital, famous apple, and Apollon Butterfly which can only be seen in Eğirdir. Kovada Lake National Park, Prostanna Antique City, Altinkum Beach, and Green Island can also be found in the area. In addition, Eğirdir's apples as used in the different dishes are confronted with a different culinary culture.

---

1 Internet connections are listed in the bibliography where you can find detailed information about the Cittaslow cities in Turkey.

4.1.3 Gökçeada (Çanakkale)

Situated in Turkey's most western point, the largest island of Gökçeada has become an important tourist attraction center in recent years, with organic products, nostalgic homes, wildlife, and alternative sports facilities. Gökçeada has been the first and only Calm Island in the world with the title of Cittaslow which was taken in June 2011. Since 2002 with organic farming activities carried out in the island and especially agricultural tourism applications carried out since 2008.

4.1.4 Gerze (Sinop)

The deep blue sea, lace-like bays and vast green, Gerze, the happiest town of Turkey, is like a magical music that you will not be able to stop listening. Having the title of Cittaslow in 2017, Gerze owes this success to the nature of its existence, to handicrafts, each of which you cannot miss, the local food – that unique and unbelievable taste, and the hospitality that warm people always welcome you with.

4.1.5 Göynük (Bolu)

Göynük is a typical Ottoman town in the western Black Sea Region, which is located between the high hills and the valleys of the rivers, at the base of the opposite slopes and on the skirts. The Mausoleum of His Holiness, Akşemseddin of Istanbul, is waiting to be visited in the garden of Gazi Süleyman Pasha Mosque. The Tower of Victory rising on the hill that dominates Göynük is also just one of the structures identified with the city. With its original architecture, local handicrafts, and traditional dishes, Göynük has an unspoilt originality in the 21st century. Göynük, which is 98 km away from the center of Bolu, attracts attention with its stubbornness and calmness to the tiring cycle of city life, despite the proximity to metropolitans such as Ankara, Istanbul, Eskişehir, and Bursa.

4.1.6 Halfeti (Şanlıurfa)

Halfeti is a small tranquil town with historical values and home to many civilizations. As a result of the Birecik Dam constructed in 2000, 3/5 part of the county was flooded. New residential areas were established for the 2/5 part. After 2000, the county became known as a hidden paradise. With the number of tourists increasing day by day, many activity areas started to form in the district. The tourists who come to the province have a chance to see the Çekem District, the Beresül (Savaşan) Village and the Rumkale (St. Nerses Church), the Baraşavma Monastery, the water cisterns, and water wells with a boat tour on the Euphrates. In this direction houses, trees, mosques, caves (Kiz cave), tea gardens, and cave cafés can be seen. Hiking trails and mountain bike trails make it possible to travel in a unique landscape full of canyons, birds, endemic plants (Black Roses), and endemic insects.

4.1.7 Perşembe (Ordu)

Perşembe District is surrounded by the Ordu centrum from the east, Ulubey from the south, Fatsa from the west, and the Black Sea from the north. The area is 226 km^2, and the county lands have a rugged appearance. The not-too-high peaks are separated by deep and steep valleys. The only untouched bays and beaches in the Black Sea region are on Perşembe, thanks to the highway passing behind Perşembe.

4.1.8 Şavşat (Artvin)

Şavşat is connected to the province of Artvin which is the easternmost of the Black Sea Region and is located to the east of Artvin province center. It is a neighbor to Ardahan. Şavşat is 71 km from the center of the province and is located at an altitude of 1,134 meters. The most important tourism and recreation areas in the province Karagöl and its surroundings are located in Meşeli Village. However, some lakes such as Arsiyan lakes, Karagöller near Bilbilan Plateau, Kazan (Samamiyet) Lakes, Akgöl, Fish and Mine are the hydro-touristic sources waiting to be evaluated in this respect.

4.1.9 Seferihisar (İzmir)

In 2009, Seferihisar joined the Cittaslow movement, which opposes the standardization of cities. Cities that are members of Cittaslow have to develop and implement projects within the framework of the criteria. Seferihisar, which is the first Cittaslow in Turkey, has fulfilled the Cittaslow criteria, which recommends local and unique cities to oppose standardization of cities by the effect of globalization. Projects such as the use of local aromatic plants in the landscape, solar-powered street lighting elements, calculation of carbon emissions, and composting plant and solar power plant construction are among Seferihisar's visionary projects. In Seferihisar, local foods can be discovered, native seeds are protected, organic agriculture is supported, and producers' markets are established where producers can sell their products without intermediaries.

4.1.10 Taraklı (Sakarya)

The area of the province is 334 km^2, of which 20 % is agriculture, 60 % is forest and shrubbery, 10 % is meadow, and 10 % is non-agricultural field. Taraklı is in the eastern part of the Marmara Region in the borders of Sakarya Province. It is 270 km from Capital Ankara. Taraklı District has a forested land structure and is established in a narrow valley. It is surrounded by high mountains and hills. Altitude from sea level is 485 meters. Despite being in the district of Marmara Region, it has a Black Sea climate.

4.1.11 Uzundere (Erzurum)

Uzundere is Turkey's 11th Cittaslow which has taken this title in Cittaslow International Coordination Committee in Italy Greve in Chanti. Uzundere has value as Turkey's highest waterfall – Tortum Falls, Tortum Lake which can host water sports such as rafting, sailing and canoeing, and OSK Monastery. Uzundere also has an important place in terms of biodiversity. Located at the western end of the Caucasus Ecological Zone, one of the world's richest bio-diversity regions, Uzundere is located in the Çoruh Valley and is home to many endemic plants, mammals, birds, and butterflies.

4.1.12 Vize (Kırklareli)

Vize is located between the two former capitals, Istanbul and Edirne, one of which entered the UNESCO heritage and one of which was prepared to enter. It is a geography that has not been discovered yet within 1.5 hours (138 km.) of Istanbul. The district, which is a paradise corner with its historical culture and its nature, is calm, charming, and a great city with its reputation in the past.

4.1.13 Yalvaç (Isparta)

Situated in the Mediterranean Region, Yalvaç is the largest district of Isparta with a population of over 20,000 in the center and 50,000 people with its villages, covering an area of 1415 km^2, 1100 meters high, on the southern skirts of the Sultan Mountains in the north-east of the Isparta Province. Despite being located in the Mediterranean Region, the terrestrial climate is observed due to its high altitude. Historically-rich Yalvaç is located side by side with Antiokheia Antique City which is one of the important centers of antiquity and has been the capital of the region.

4.1.14 Yenipazar (Aydın)

Yenipazar is a district center located in Aegean Region, 40 km away from Aydın Province. It is located in the middle of the Büyük Menderes Basin with a surface area of 180 km^2, at the foot of Madran Baba Mountain covered with forests to the north.

5 Conclusion

With the process of globalization and modernization, mankind, who has become increasingly isolated from the crowd, finds itself in a fast life that has consumed more with the effect of capitalism and away from the natural environment and isolated from the moral values. This individuation has made people selfish by transforming them into *"those who think only of themselves or who overstep their own interests over the interests of everyone else"*. All these transformations have not brought happiness to human beings. Nowadays, some people think that happiness can be recovered not by thinking about themselves first, but by new systems where everyone thinks of each other and the community itself. In order for such systems to exist, it has been noticed that modern city experiences, which are focused on fast consumption and fast life, must be transformed.

For a more sustainable world, some consumers are aware of the need to exist within new sustainable consumption systems. In this process, with the formation of the green consumer concept, they realized that people need to reach the process of simplification, which only consumes as much as they need, recycling old things, prolonging the period of use, confine with as few as possible and shaping all lifestyles around it. This simplification process brought with it the desire to live more slowly. The philosophy of eating slower and better-quality food and preserving traditional tastes has led to the concept of Cittaslow with this philosophy in tiny and protected areas, with the vigor of preserving local trades and cultural heritage.

This movement, which first appeared in Italy in 1999, spread to the whole world with the establishment of new Cittaslows by applying to the International Cittaslow Union with the provision of certain determined criteria. Nowadays, 236 cities in 30 countries / regions have been accepted as Cittaslow. Turkey has 14 Cittaslow cities. Turkey is the 4th country with the most Cittaslow cities in Europe, after Italy (83 Cittaslows) which is the founder country of Cittaslow movement, Poland (27 Cittaslows), and Germany (17 Cittaslows).

In general, Cittaslow cities are places where sustainable consumption practices exist, where sustainable regional development and sustainable tourism are targeted, where local, cultural, historical heritage is protected, local craftsmen and traditions are protected, people are able to socialize and socialize in a shared environment, clean, good, fair food and production processes exist, usually boutique areas with a population of less than 50,000 people.

Cittaslows, in addition to all these goals, provide the opportunity for people to obtain maximum satisfaction from their lives with a slower life by providing protection from the disadvantages of globalization, and they are at the beginning of a new lifestyle practice.

References

Assadourian, E. (2010). "The rise and fall of consumer cultures". *State of the World 2010: Transforming Cultures.* Worldwatch Institute Report. New York: W. W. Norton and Co.

Çakır, G. & Çakır, A (2015). *"Slow Food" Bilinçli Mutfak, "Trakya Bölgesindeki Yiyecek ve İçecek Mesleki Eğitim Okullarında Kalitenin Arttırılması" Projesi.* Ankara, s. 53–79, www.trakyagastronomi.com/slow-food-hareketi-nedir.html, (08.06.2018).

Cittaslow (2018a). http://cittaslowturkiye.org/#cittaslow, (08.06.2018).

Cittaslow (2018b). http://www.cittaslow.org/, (08.06.2018).

Cittaslow (2018c). http://cittaslowturkiye.org/uyelik-sureci-ve-kriterler/, (08.06.2018).

Cittaslow (2018d). http://cittaslowturkiye.org/, (10.06.2018).

CittaslowOfficial Website. (2018). http://cittaslowturkiye.org/cittaslow-akyaka/, (10.06.2018).

CittaslowOfficial Website. (2018). http://cittaslowturkiye.org/cittaslow-egirdir/, (10.06.2018).

CittaslowOfficial Website. (2018). http://cittaslowturkiye.org/cittaslow-gokceada/, (10.06.2018).

Cittaslow Official Website. (2018). http://cittaslowturkiye.org/cittaslow-gerze/, (10.06.2018).

Cittaslow Official Website. (2018). http://cittaslowturkiye.org/cittaslow-goynuk/, (10.06.2018).

Cittaslow Official Website. (2018). http://cittaslowturkiye.org/cittaslow-halfeti/, (10.06.2018).

Cittaslow Official Website. (2018). http://cittaslowturkiye.org/cittaslow-persembe/, (10.06.2018).

Cittaslow Official Website. (2018). http://cittaslowturkiye.org/cittaslow-savsat/, (10.06.2018).

Cittaslow Official Website. (2018). http://cittaslowturkiye.org/cittaslow-seferihisar/, (10.06.2018).

Cittaslow Official Website. (2018). http://cittaslowturkiye.org/cittaslow-tarakli/, (10.06.2018).

CittaslowOfficial Website. (2018). http://cittaslowturkiye.org/cittaslow-uzundere/, (10.06.2018).

CittaslowOfficial Website. (2018). http://cittaslowturkiye.org/cittaslow-vize/, (10.06.2018).

CittaslowOfficial Website. (2018). http://cittaslowturkiye.org/cittaslow-yalvac/, (10.06.2018).

Cittaslow Official Website. (2018). http://cittaslowturkiye.org/cittaslow-yenipazar/, (10.06.2018).

Craig-Lees, M. & Hill, C. (2002). "Understanding Voluntary Simplifiers". *Psychology&Marketing*, 19(2): 187-210.

Elgin, D. ve Mitchell, A. (1977). "Voluntary Simplicity: Lifestyle of the Future?".*The Futurist*, 11(4): 200-261.

Gerzema, J. (2010). *Spend Shift: How the Post Crisis Values Revolution Is Changing the Way We Live, Shop and Buy*. San Francisco: Jossey-Bass.

Global Footprint Network. (2009). "*How we can bend the curve: Trending toward a sustainable future*". Global Footprint Network Annual Report, http://www.footprintnetwork.org/images/uploads/Global_Footprint_Network_2009_annual_report.pdf., (08.06.2018).

Güven, E. (2011). "Yavaş Güzeldir: "Yavaş Yemek" ten "Yavaş Medya"ya Hızlı Tüketime Dair Bir Çözüm Önerisi", *Selçuk İletişim Dergisi*, 7: 113-121.

Kasulis, J. J., Huettner, D. A. & Dikeman, N. J. (1981). "The Feasibility of Changing Electricity Consumption Patterns". *Journal of Consumer Research*, 8(3): 279-290.

Kilbourne, W.E., Beckmann, S. C. & Thelen, E. (2002)."The Role of The Dominant Social Paradigm in Environmental Attitudes -A Multinational Examination". *Journal of Business Research*, 55(3): 193-204.

Kotler, P. (2011). "Reinventing Marketing To Manage The Environmental Imperative", *Journal of Marketing*, 75(4): 132-135.

Leary, R. B., Vann, R. J., Mittelstaedt, J. D., Murphy, P. E. & Sherry, J. F. (2014). "Changing the Marketplace One Behavior at a Time: Perceived Marketplace Influence and Sustainable Consumption". *Journal of Business Research*, 67(9): 1953-1958.

Mcdonald, S., Oates, C. J., Young, C. W. & Hwang, K. (2006). "Toward Sustainable Consumption: Researching Voluntary Simplifiers". *Psychology & Marketing*, 23(6): 515-534.

Myers, N. (2000). "Sustainable Consumption". *Science*, 287(5462): 2419.

Odabaşı, Y. (2013). *Tüketim Kültürü*. İstanbul: Sistem Yayıncılık.

Quelch, J. A. & Jocz, K. E. (2007). *Greater Good: How Good Marketing Makes for a Better World*. Boston: Harvard Business Press.

Saad, L. (2006). "*Americans see environment as getting worse*",http://www.gallup.com/, (01.06.2016).

Schor, J. B. (1999). *The Overspent American: Why We Want What We Don't Need.* New York: Harper Perennial.

Sezgin, M. & Ünüvar, Ş. (2011). *Yavaş Şehir. Sürdürülebilirlik ve Şehir Pazarlaması Ekseninde.* Konya: Çizgi Kitabevi.

Shama, A. (1996). "A comment on The meaning and morality of voluntary simplicity: History and hypothesis on deliberately denied materialism", *Consumption and Marketing: Macrodimensions* (Ed.) R.W., Belk, N., Dholakia, A., Venkatesh,Cincinnati: South-Western College Publishing.

Sharma, A., Iyer, G. R., Mehrotra, A. & Krishnan R. (2010). "Sustainability and Business-To-Business Marketing: A Framework and Implications". *Industrial Marketing Management,* 39(2): 330–341.

Sheth, J. N., Sethia, N. K. & Srinivas, S. (2011). "Mindful Consumption: A Customer-Centric Approach to Sustainability". *Journal of The Academy of Marketing Science,* 39(1): 21–39.

Sırım, V. (2012). "Çevreyle Bütünleşmiş Bir Yerel Yönetim Örneği Olarak "Sakin Şehir" Hareketi ve Türkiye'nin Potansiyeli". *Tarih Kültür ve Sanat Araştırmaları Dergisi,* 1(4): 119–131.

Slow Food (2018a). http://www.slowfood.com/international/2/our-philosophy, (08.06.2018).

Slow Food (2018b). https://www.slowfood.com/about-us/our-history/, (08.06.2018).

Strizhakova, Y. & Coulter, R. A. (2013). "The "Green" Side of Materialism in Emerging BRIC and Developed Markets: The Moderating Role of Global Cultural Identity". *International Journal of Research in Marketing,* 30(1): 69–82.

Yurtseven, R., Kaya, O. & Harman, S. (2010). *Yavaş Hareketi.* Ankara: Detay Yayıncılık.

Zawadzka, A. K. (2017). "Making Small Towns Visible in Europe: The Case of Cittaslow Network–The Strategy Based on Sustainable Development". *Transylvanian Review of Administrative Sciences,* 13(SI): 90–106.

Serap Pelin Türkoğlu and Yasemin Hancıoğlu
Estimation of Countries' Development Status with Logistic Regression Analysis

1 Introduction

The concept of human capital (Taban & Kar, 2006: 163), which involves the knowledge, ability, skill, social status, health status and education level of the individual within a given society, is recognized as the driving force behind economic growth and development (Özyakışır, 2011: 64; Karataş & Çankaya, 2010).

Economic growth is measured by the change in national income over a period of time, whereas general development and progress are measured on the basis of the positive changes undertaken by a country, in terms of their social, economic and political aspects in addition to their economic growth (Erkekoğlu, 2007: 28). "Development" is often taken to mean rising incomes. A still common view equates development with growth in average income, though there has been a shift in emphasis since the 1970s to a focus on the distribution of incomes (Anand & Ravallion, 1993: 133).

A country's wealth in terms of human capital is related to certain factors, like health and education, which serve to improve labour productivity. In order to ensure continuity in economic development, individuals need to be equipped with knowledge capable of improving their performances; in other words, human capital needs to be educated. Countries like Taiwan, South Korea and Singapore, which are developing fast, are placing importance on primary and secondary education, and they have increased the duration of compulsory education to nine years. Another important feature of these countries that has been attracting attention is their strong focus on higher education. From these examples, it can be argued that education is a primary element of human capital accumulation and as such, plays a critical role in economic development (Eser & Ekiz Gökmen, 2009: 46).

The ability of emerging countries to achieve economic development depends primarily on their human capital development (Keskin, 2011: 125). Development of human capital is a prominent issue in the area of social services. The role of social services – particularly basic health and education – started to receive greater attention in the 1980s, although these services have been viewed mainly as instruments for raising the incomes of the poor. In all these approaches, income growth of one sort or another is what development is all about (Anand & Ravallion, 1993: 133).

In the second chapter, the Human Development Index and its indicators were discussed, while in the third chapter, studies examining the relationship between the Human Development Index and economic development were reviewed. In the following chapter, the dynamics affecting the development status of countries in the Human Development Index are analyzed using logistic regression, after which the results of the study are discussed, followed by a general evaluation.

2 Human Development Index

Human development is defined as the enlargement of the possibilities and options available to people in their pursuit of securing the standard of living they desire (Sezgin Nartgün, 2013: 81; Tunç & Ertuna, 2015: 155). In other words, human development is a process of enlarging people's choices, that is, as they acquire more capabilities, they will be able to enjoy more opportunities to use those capabilities. However, human development is also the objective, so it is both a process and an outcome (UNDP, 2015: 1–272).

A rather different view of the meaning of development (progress) recently emerged in the 1990 Human Development Report (HDR) produced by the United Nations Development Programme (UNDP). The essence of this view is that human development – what people can actually do and be – is the primary goal of economic development (Anand & Ravallion, 1993: 133–134).

The Human Development Index (HDI) (Çivi et al., 2008: 16; McGillivray & White, 1993: 183), which has been published by the United Nations Development Programme (UNDP) Office in the United Nations (UN) since 1990, is based on the premise that the development of living standards cannot be explained by economic indicators alone. The HDI includes health, education and income indicators, as well as prominent current and economic issues of the respective year, and is part of the Human Development Reports (Hacıoğlu Deniz & Haykır Hobikoğlu, 2012: 123, Hicks, 1997: 1283).

The HDI is a composite index that focuses on three basic dimensions of human development: the likelihood of leading a long and healthy life, measured by life expectancy at birth; the ability to acquire knowledge, measured by mean years of schooling and expected years of schooling and the ability to achieve a decent standard of living, measured by gross national income per capita. The HDI has an upper limit of 1.0 (UNDP, 2015: 1–272).

In the United Nations Development Reports, countries are classified as low, medium, high and very high, in terms of human development, based on their

HDI scores. As a tool used for ranking the life quality of different countries, the HDI index has become increasingly popular (Engineer et al., 2008: 172), functioning as an important alternative to the traditional unidimensional development index (Sagar & Najam, 1998: 249).

3 Literature

There are many studies in the literature that have investigated the relationship between the Human Development Index, human capital and economic development, using various statistical methods and human capital indicators. Some of the most relevant studies include the following:

Kumar (1991) constructed the Human Development Index for 17 Indian States, ranking these states by calculating the human development score according to the methodology specified for each state in the Human Development report.

McGillivray (1991) examined both the composition and usefulness of the Human Development Index as a composite development indicator using correlation analysis. Results obtained from the analysis suggested that there was a problem in the composition of the Human Development Index published in 1990 and that the indicators used for determining intercountry development levels, such as gross domestic product (GDP), failed to adequately explain human development.

Benhabib and Spiegel (1994) examined the relationship between the Cobb-Douglas production function and economic development by comparatively analyzing physical capital and human capital stocks of countries. The results of the analyses revealed that human capital was an insignificant input for explaining the growth rates per capita.

Gormely (1995) argued that the United Nations Development Programme's method of estimating the contribution of per-capita income to human development is inappropriate and leads to misleading Human Development Index country rankings. For countries below the world-average per-capita income, the United Nations Development Programme's estimation formula allows per-capita income, as measured by each country's Human Development Index. However, for countries above the world-average per-capita income, high incomes, as measured by the Human Development Index, are estimated to make no additional contribution to their human development. In the study, an alternative treatment of income was illustrated, and it showed that there was an appreciable alteration to the country Human Development Index rankings as a result of this modification.

Anand and Sen (2000) aimed to integrate the concern for human development in the present with that in the future. For sustainable human development, they argued the notion of ethical "universalism" applied within and between generations. Economic sustainability is often seen as a matter of intergenerational equity, but the specification of what is to be sustained is not always straightforward. Therefore, Anand and Sen sought to explore the relationship between distributional equity, sustainable development, optimal growth and pure time preference.

Ranis and Stewart (2000) stated that the relationship between economic growth and human development linked two chains. Comparative regressions showed that there was a significant relationship between these chains (public expenditure in education and health). These expenditures are especially important in the chain that stretches from human development to economic growth, while investment rate and income distribution are important in the chain that stretches from economic growth to human development.

Crafts (2002) revised the Human Development Index calculations to compare 1870, 1913, 1950 and 1999. For this, the recently modified formula of the Human Development Index and the latest available data for GDP per capita were used. The results indicated that the Human Development Index score in most of today's less-developed countries exceeded that of Western Europe in 1870.

Çakmak and Gümüş (2005) tested the hypothesis, "There is a long-term positive relationship between human capital and economic growth" by employing a "Co-Integration Analysis" for the period between the years 1960 and 2002 using an index formed by applying different weights to primary, secondary and higher education graduates. The results indicated that there was a positive relationship between human capital and economic growth in Turkey.

Taban and Kar (2006) investigated the causal relationship between human capital, whose contribution to economic growth is positive in endogenous growth theories, and economic growth for the period between 1969 and 2001 by using Turkish data. In order to test this relationship, the Human Development Index, Education Index, Combined Enrolment Ratio, Life Expectancy Index and GDP, as an indicator of economic growth, were used in the study. The results showed there to be an interaction between economic growth and human capital variables.

Altay and Pazarlıoğlu (2007) examined the relationship between competitive power and human capital indicators using 2000–2004 data on the first 50 countries in the international competition power ranking. The model results showed the effect of human capital on macroeconomic indicators and competitive power, finding specifically that the education variable could change a country's international competition ranking.

Acar Bolat and Arıcıgil Çilan (2007) aimed to identify indicators capable of determining development discrepancy in developing countries from regions in Europe, Central Asia, the Middle East and Africa, in terms of the components of the Human Development Index. In order to determine these discrepancies, the discriminant analysis method was used. Results showed that the indexes that differentiate the regions of the developing countries are education and GDP indices.

Öz et al. (2009) compared Turkey with the EU countries in terms of 23 different human capital indicators, including education, health and labour force, to name a few. Results from the cluster analysis revealed that Turkey did not resemble old or new members of the EU country in any of the three name areas (education, health and labour force).

Keskin (2011) sought to explain the relationship between human capital and economic growth by using a multi-linear regression model on the data of 177 member countries of the United Nations. The study revealed that human capital was very important for economic development. Furthermore, regression analysis results showed that education level, literacy rate, R & D expenditures and public health expenditures had significant effects on economic development.

Demiray Erol (2013) aimed to determine the levels of development of countries by comparing socio-economic indicators of the EU member states and Turkey. In the study, socio-economic development indicators of the EU member states and Turkey were calculated by using the principal component analysis method. Nineteen variables, including life expectancy at birth, urban population ratio, health expenditures, per capita income, inflation rate, the share of expenditure on education in GDP, employment rate, R & D expenditures and crime rate were used to explain the socio-economic development levels. The analysis results showed that Turkey was the least developed country out of the EU countries in terms of social development.

Koç (2013) tested the impact of human capital on economic growth for EU countries by using 2012 data and cross-sectional analysis method. Results of the analysis showed that human capital had a statistically significant and positive impact on economic growth.

Sezgin Nartgün et al. (2013) investigated whether there were differences in terms of human development criteria, education index and GDP among the European Union states, the EU candidate states and the EU potential candidate states by using the document review method. Results of the review showed that the index values of Turkey were lower than those of the EU states, the EU candidate states, and the EU potential candidate states.

Bal et al. (2014) aimed to determine the causal relationship between the human capital index and economic growth in Turkey and BRICS (Brazil, Russia, India, China and South Africa) countries by using Panel Data Analysis for between the years 1995–2011. According to the results, the long-term relationship between human capital and economic growth was significant and positive.

Başar et al. (2015) measured the relative efficiencies of each country that was grouped as very high, high, medium and low human development, as to their Human Development Index values, by evaluating each group separately. As a result of the data envelopment analysis model, it was determined that the countries in the first rank in their groups according to the Human Development Index were for the most part efficient, but there were some exceptions.

Santos et al. (2017) estimated the Human Development Index for Latin American countries in 2013 and 2014 by using data mining techniques. Estimates of the models did not show statistically significant differences from the trend of the Human Development Index announced in the last report of the United Nations Development Programme.

4 Analysis of Factors Affecting the Development Status of Countries in the Human Development Index

4.1 Data, Aim, Method

The aim of this study was to identify the factors affecting the development status of the selected countries, to determine the impact weights of these factors and to estimate the development status of the countries. To achieve this aim, 105 countries listed on the Human Development Index were examined. The countries that were within the very high and high human development groups were selected, with the country data for 2014 and the data set from the 2015 Human Development Report being used.

Logistic regression analysis is a statistical method used when the dependent variable is categorical (patient / healthy, bankrupt / non-bankrupt, etc.). Independent variables in logistic regression analysis can be categorical, continuous or discrete. This method can be classified according to the category number of the dependent variable, that is, if the category number of the dependent variable is two, binary logistic regression is applied, while if the category number is more than two, multi-nominal logistic regression is applied.

One of the objectives of the logistic regression analysis is to classify individuals into different groups and to investigate the relationship between dependent and independent variables (Çokluk et al., 2016: 50). In the logistic regression analysis, the assumptions that independent variables have normal distribution and their covariances are the same on a group basis are not required (Kalaycı, 2010: 273). The logistic regression model is expressed by the following equation:

$$L = \ln\left(\frac{P_i}{1-P_i}\right) = b_o + \sum_{i=1}^{n} b_i X_i \quad (1)$$

where "p_i" denotes a value for the predicted probability of an event's occurrence, "$1-p_i$" denotes the predicted probability of an event's non-occurrence, and "b_o" and "b_i" are the coefficients of the constant term and independent variables, respectively. The equation showing the probability of "p_i", which lies between 0 and 1, is presented just below:

$$p_i = \frac{1}{1+e^{-\left(b_o + \sum_{i=1}^{n} b_i X_i\right)}} \qquad (2)$$

The odds ratio, also called the strength of the association, represents the odds that an outcome will occur given a particular exposure, compared to the odds of the outcome occurring in the absence of that exposure. The odds ratio is important for interpretation of the logistic regression model. The natural logarithm of the odds ratio gives the logit value (L), which lies between $-\infty$ and $+\infty$.

$$Odds = \frac{p_i}{1-p_i} \qquad (3)$$

The maximum likelihood method is used for the estimation of parameters in logistic regression analysis. This method is applied to find the logistic regression coefficients that maximize the likelihood function (Çokluk et al., 2016: 62). In order to test the significance of the model, -2LogL, developed model chi-square and baseline model chi-square statistics were used. Furthermore, in the logistic regression analysis, the measurement of the relationship between dependent and independent variables was made by Cox & Snell R^2 and Nagelkerke R^2 statistics.

Analyses were done using the binary logistic regression model. Since the dependent variable in the logistic regression analysis was categorical, the countries, which were considered as the dependent variables of the study, were classified as either developed or developing. The classification of the selected countries as developed or developing countries was based on their status of being either above or below the average of the per capita gross national income. In other words, if the average gross national income per capita of the selected country was above US $ 26,949 (developed), the dependent variable was "1", whereas if it was below this figure (developing), the dependent variable was "0".

Life expectancy at birth (LEB), expected years of schooling (EYS) and mean years of schooling (MYS) variables were used as the independent variables in the study. The Human Development Index (HDI) scores were intended to be considered as independent variables. However, these variables were excluded from the analysis because they had a high correlation with other independent variables.

Tab. 1: Descriptive Statistics for Independent Variables.

Statistics Type	LEB (Year)	EYS (Year)	MYS (Year)
Mean	76.9	14.9	10.2
Standard Deviation	3.8	1.7	1.8
Median	76.3	14.7	10.3
Minimum Value	69.4	11.8	5.8
Maximum Value	84.0	20.2	13.1

The descriptive statistics for the independent variables of the study are given below in Tab. 1.

According to Tab. 1, the variable with the highest mean was "LEB", while the variable with the lowest mean was "MYS". The variable with the maximum value was "LEB", while the variable with the minimum value was "MYS".

4.2 Logistic Regression Analysis Results

In order to determine the factors affecting the development status of the selected 105 countries, binary logistic regression was applied. The SPSS 16.0 software package was used in the logistic regression analysis, for which the enter method was applied. In this method, all independent variables in a block were entered. The significance level (α value) used for the study was 0.05.

The logistic regression analysis is not a method that requires many assumptions. However, prior to conducting the analysis, it is necessary to check whether there is a high correlation between outliers and variables. Outliers were identified by applying Mahalanobis distance of independent variables, which revealed that there were no outliers. A correlation analysis was conducted to determine the relationship between the independent variables. The correlation analysis table for the independent variables (HDI, LEB, EYS, MYS) is given below.

According to the correlation analysis results presented in Tab. 2, there was a positive, significant relationship between the independent variables. In addition, there was a high correlation between the "HDI" variable and other independent variables. In order to prevent the multicollinearity problem, the "HDI" variable was not included in the analysis.

When the Omnibus tests of model coefficients in Tab. 3 are examined, the fact that the p value related to the chi-square test of the model was significant shows the existence of a relationship between dependent and independent variables. Moreover, the fact that this value was significant means the rejection of the null

Tab. 2: Correlations between Independent Variables.

		HDI	LEB	EYS	MYS
HDI	Correlation Coefficient	1	0.787**	0.753**	0.678**
	P		0.000	0.000	0.000
LEB	Correlation Coefficient	0.787**	1	0.522**	0.340**
	P	0.000		0.000	0.000
EYS	Correlation Coefficient	0.753**	0.522**	1	0.495**
	P	0.000	0.000		0.000
MYS	Correlation Coefficient	0.678**	0.340**	0.495**	1
	P	0.000	0.000	0.000	

**: Correlation is significant at the 0.01 level.

Tab. 3: Omnibus Tests of Model Coefficients.

		Chi-square	df	p
Step 1	Step	69.883	3	0.000
	Block	69.883	3	0.000
	Model	69.883	3	0.000

hypothesis, namely, that there is no difference between the baseline model, which includes only a constant term, and the new model, which includes explanatory variables.

In Tab. 4, which summarizes the developed model, Cox & Snell R^2 and Nagelkerke R^2 statistics are presented. These statistics indicate the degree of relationship between the dependent variable and the independent variables, which in this case revealed that there was a 49 % and 67 % relationship between the dependent variable and the independent variables, according to the Cox & Snell R^2 and Nagelkerke R^2 statistics, respectively. The -2 log likelihood (-2LL) is a statistic that expresses the fit of the model, where the smallest value of -2LL, "0", indicates a perfect fit.

In Tab. 5, the result of the Hosmer & Lemeshow test, also referred to as the chi-square goodness of fit test, showed there to be no significance (0.666>0.05), which indicates that the model data fit was sufficient. In other words, the expected values were not different from the observed values.

The Classification Table obtained from the logistic regression analysis is presented in Tab. 6. According to this table, 61 of the 67 developing countries were classified correctly and six of them were misclassified, making the percentage of

Tab. 4: Model Summary.

Step	-2 Log likelihood	Cox & Snell R^2	Nagelkerke R^2
1	67.563	0.486	0.666

Tab. 5: Hosmer and Lemeshow Test.

Step	Chi-square	df	p
1	7.453	8	0.489

Tab. 6: Classification Table.

	Observed		Predicted		Percentage Correct
			Developed		
			0	1	
Step 1	Developed	0	61	6	91.0
		1	8	30	78.9
	Overall Percentage Correct				86.7

correct classification to be 91 %. While 30 of 38 developed countries were classified correctly, eight of them were misclassified, making the percentage of correct classification to be 78.9 %. The percentage of overall correct classification of the developed model was 86.7 %.

Tab. 7 shows the variables included in the developed model and the regression coefficients. The only variable that had a positive, significant effect on the development status of the selected countries was "LEB". The other independent variables had no significant effect on the development status of the selected countries. A one-unit increase in the LEB variable led to an increase of 1.844 in the country's development advantage (odds value).

Using the values presented in Tab. 7, the logistic regression model was created on the basis of the following equation. With this model, the probability that countries are either developed or developing can be estimated.

$$\ln(\text{odds}) = -51.092 + 0.612 \text{ LEB} + 0.072 \text{EYS} + 0.181 \text{MYS} \quad (4)$$

In looking at the results of the logistic regression analysis, it is clear that the variable, "LEB", positively and significantly affected the development status of the selected countries. The "EYS" and "MYS" variables, on the other hand, had no significant effect on the development status of the countries.

Tab. 7: Regression Coefficients.

		B	Standard Error	Wald	df	p	Exp(B)
Step 1	LEB	0.612	0.133	21.064	1	0.000	1.844
	EYS	0.072	0.241	0.088	1	0.766	1.074
	MYS	0.181	0.219	0.684	1	0.408	1.199
	Constant	-51.092	9.400	29.543	1	0.000	0.000

5 Conclusion

The effect of human capital, which is defined as the sum of values, such as skill, knowledge, dynamism and experience, that the labour force has in the production process, in terms of using production factors more efficiently, on economic development was first considered in the 1960s. In examining the human capital concept, it is clear that education and health are two important factors, which, in their function within the human capital concept, are in a mutually supportive relationship.

The idea of human capital accumulation has long been recognized as an important factor affecting economic development (Benhabib & Spiegel, 1994: 166). In order to increase the development status of a country, more resources should be allocated to education and health, and this should be viewed as an investment. The higher the health and education indicators of a country, the more likely there will be economic development.

Regarding less developed and developed countries, increasing human capital and using it effectively are very important. In order to increase human capital accumulation, a society formed of healthy, educated individuals is needed. Moreover, to attain the desired productivity from the human capital, the satisfaction of the individuals, in terms of social needs and financial opportunities, is an indispensable condition (Eser &Ekiz Gökmen, 2009: 42).

Results of the logistic regression analysis showed that 86.7 % of the 105 countries in the Human Development Index were correctly classified. Furthermore, it was determined that the prediction of the developed model was consistent with the data, and that the "LEB" variable had a positive significant effect on determining the probabilities of the development status of the selected countries.

In the analysis of the factors affecting the development status of the selected countries, the logistic regression analysis produced significant results. In addition, in estimating the probabilities of the development status of the countries, the developed logistic regression model proved to be an effective tool, one that could be used by decision makers to shed light on their countries' development prospects.

References

Acar Bolat, B. & Arıcıgil Çilan, Ç. (2007). "İnsani Gelişme İndeksi Bileşenleri Açısından Gelişmekte Olan Ülkelerin Diskriminant Analizi ile Karşılaştırıl ması", 38. *Uluslararası Asya ve Kuzey Afrika Çalışmaları Kongresi ICANAS*, 10–15 Eylül 2007, Ankara.

Altay, A. & Pazarlıoğlu, M.V. (2007). "Uluslararası Rekabet Gücünde Beşeri Sermaye: Ekonometrik Yaklaşım", *Selçuk Üniversitesi Karaman İİBF Dergisi*, 9(12): 96–108.

Anand, S. & Ravallion, M. (1993). "Human Development in Poor Countries: On the Role of Private Incomes and Public Services", *Journal of Economic Perspectives*, 7(1): 133–150.

Anand, S. & Sen, A. (2000). "Human Development and Economic Sustainability", *World Development*, 28(12): 2029–2049.

Bal, H., Algan, N., Manga, M. & Kandır, E. (2014). "Beşeri Sermaye ve Ekonomik Büyüme İlişkisi: BRICS Ülkeleri ve Türkiye Örneği", *International Conference on Eurasian Economies*, 1–3 Temmuz, Makedonya.

Başar, S., Eren, M. & Eren, M. (2015). "Ülkelerin İnsani Gelişmişlik Endeksi Değişkenlerine Göre Etkinliklerinin İncelenmesi", *International Conference on Eurasian Economies*, 9–11 Eylül, Rusya.

Benhabib, J. & Spiegel, M. M. (1994). "The Role of Human Capital in Economic Development Evidence from Aggregate Cross-Country Data", *Journal of Monetary Economics*, 34(2): 143–173.

Crafts, N., (2002). "The Human Development Index 1870-1999: Some Revised Estimates", *European Review of Economic History*, 6(3): 395–405.

Çakmak, E. & Gümüş, S. (2005). "Türkiye'de Beşeri Sermaye ve Ekonomik Büyüme: Ekonometrik Bir Analiz (1960-2002)", *Ankara Üniversitesi SBF Dergisi*, 60(1): 59–72.

Çivi, E., Erol, İ., İnanlı, T. & Erol, E.D. (2008). "Uluslararası Rekabet Gücüne Farklı Bakışlar", *Ekonomik ve Sosyal Araştırmalar Dergisi*, 4(1): 1–22.

Çokluk, Ö., Şekercioğlu, G. & Büyüköztürk, Ş. (2016). *Sosyal Bilimler için Çok Değişkenli İstatistik SPSS ve LISREL Uygulamaları*, Ankara: Pegem Akademi.

Demiray Erol, E. (2013). "Türkiye ve Avrupa Birliği Üyesi Ülkelerin Sosyo-Ekonomik Gelişmişlik Düzeylerinin Karşılaştırmalı Analizi", *Sosyal ve Beşeri Bilimler Dergisi*, 5(1): 198–208.

dos Santos, C. B., Pedroso, B., Guimaraes, A. M., Carvalho, D. R., & Pilatti, L. A. (2017). "Forecasting of Human Development Index of Latin American Countries Through Data Mining Techniques", *IEEE Latin America Transactions*, 15(9): 1747–1753.

Engineer, M., King, I. & Roy, N. (2008). "The Human Development Index as a Criterian for Optimal Planning", *Indian Growth and Development Review*, 1(2): 172–192.

Erkekoğlu, H. (2007). "AB'ye Tam Üyelik Sürecinde Türkiye'nin Üye Ülkeler Karşısındaki Göreli Gelişme Düzeyi: Çok Değişkenli İstatistiksel Bir Analiz", *Kocaeli Üniversitesi Sosyal Bilimler Enstitüsü Dergisi*, 14(2): 28–50.

Eser, K. & Ekiz Gökmen, Ç. (2009). "Beşeri Sermayenin Ekonomik Gelişme Üzerindeki Etkileri: Dünya Deneyimi ve Türkiye Üzerine Gözlemler", *Sosyal ve Beşeri Bilimler Dergisi*, 1(2): 41–56.

Gormely, P. J. (1995). "The Human Development Index in 1994: Impact of Income on Country Rank", *Journal of Economic and Social Measurement*, 21(4): 253–267.

Hacıoğlu Deniz, M. & Haykır Hobikoğlu, E. (2012). "Cinsiyete Göre Gelişme Endeksi Çerçevesinde Kadın İstihdamının Ekonomik Değerlendirilmesi: Türkiye Örneği", *International Conference on Eurasian Economies*, 11–13 Ekim, Kazakistan.

Hicks, D. A. (1997). "The Inequality-Adjusted Human Development Index: A Constructive Proposal", *World Development*, 25(8): 1283–1298.

Kalaycı, Ş. (2010). *SPSS Uygulamalı Çok Değişkenli İstatistik Teknikleri*, Ankara: Asil Yayın Dağıtım.

Karataş, M. & Çankaya, E. (2010). "İktisadi Kalkınma Sürecinde Beşeri Sermayeye İlişkin Bir Inceleme", *Mehmet Akif Ersoy Üniversitesi Sosyal Bilimler Enstitüsü Dergisi*, 2(3): 29–55.

Keskin, A. (2011). "Ekonomik Kalkınmada Beşeri Sermayenin Rolü ve Türkiye", *Atatürk Üniversitesi İktisadi ve İdari Bilimler Dergisi*, 25(3–4): 125–158.

Koç, A. (2013). "Beşeri Sermaye ve Ekonomik Büyüme İlişkisi: Yatay Kesit Analizi ile AB Ülkeleri Üzerine Bir Değerlendirme", *Maliye Dergisi*, 165: 241–258.

Kumar, A. K. S. (1991). "UNDP's Human Development Index A Computation for Indian States", *Economic and Political Weekly*, 26(41): 2343–2345.

McGillivray, M. (1991). "The Human Development Index: Yet Another Redundant Composite Development Indicator?", *World Development*, 19(10): 1461–1468.

McGillivray, M. & White, H. (1993). "Measuring Development? The UNDP's Human Development Index", *Journal of International Development*, 5(2): 183–192.

Öz, B., Taban, S. & Kar, M. (2009). "Kümelenme Analizi ile Türkiye ve AB Ülkelerinin Beşeri Sermaye Göstergeleri Açısından Karşılaştırılması", *Eskişehir Osmangazi Üniversitesi Sosyal Bilimler Dergisi*, 10(1): 1–30.

Özyakışır, D. (2011). "Beşeri Sermayenin Ekonomik Kalkınma Sürecindeki Rolü: Teorik Bir Değerlendirme", *Girişimcilik ve Kalkınma Dergisi*, 6(1): 46-71.

Ranis, G. & Stewart, F. (2000). "Economic Growth and Human Development", *World Development*, 28(2): 197-219.

Sagar, A. D. & Najam, A. (1998). "The Human Development Index: A Critical Review", *Ecological Economics*, 25(3): 249-264.

Sezgin Nartgün, Ş., Akın Kösterelioğlu, M. & Sipahioğlu, M. (2013). "İnsani Gelişim İndeksi Göstergeleri Açısından AB Üyesi ve AB Üyeliğine Aday Ülkelerin Karşılaştırılması", *Trakya Üniversitesi Eğitim Fakültesi Dergisi*, 3(1): 80-89.

Taban, S. & Kar, M. (2006). "Beşeri Sermaye ve Ekonomik Büyüme: Nedensellik Analizi, 1969-2001", *Sosyal Bilimler Dergisi*, 1: 159-182.

Tunç, O. & Ertuna, Ö. (2015). "İnsani Gelişme Endeksi Türkiye Simülasyonu ve Seçilmiş Ülkelerle Karşılaştırılması", *Journal of Management, Marketing and Logistics*, 2(2): 132-157.

UNDP, (2015). *Human Development Report 2015*, Work for Human Development, ISSN: 0969-4501, New York.

Hüseyin Yılmaz and Abdullah Elmas

The Study on Turkey's Demographic Window of Opportunity

1 Introduction

In short, the demography can be defined as "Population Science" and it is not only the numerical size of the population, but also a science that examines the basic variables of the population such as birth, death, migration and their sub-concepts such as marriage, divorce and education. This science is closely related to many disciplines ranging from geography to biology, from medicine to sociology, from politics to military, from health services to education system and economics. The emergence of the demography, which spreads such a wide area, as a scientific discipline has been attributed to John Graunt's work *"Natural and Political Observations on the Bills of Mortality* published in 1662." Malthus's *An Essay on the Principle of Population* was the first book in the field of demography. Researchers' interest over this science has increased following these studies, and demographic data have played an important role in shaping the policies of states.

Various population policies have been applied from period in ancient times (a person per 2 km^2) to the present days (20 thousand people per km^2). The idea that population growth is beneficial underlines at the basis of these demographic policies. The population policies are applied in terms of not only the socio-economic structure of the countries but also the conditions of the period. According to the period of the conditions in Turkey, demographic policies are applied to increase the population between 1923 and 1965, to reduce population between 1965 and 2007 and to enhance the quality of the population after 2007.

Until the 18th century, the population grew very slowly due to the wars and the inadequacy of health services. In the 20th century, the development of health services and the rapid advancement of technology enabled to increase the average life expectancy. As a result of this situation, states faced the problem of elderly population that they have never experienced before. This situation revealed the importance of high proportion of the young population and emerged the concept of "The Demographic Window of Opportunity", which is defined as an economic bounce by taking advantages of the young population. Every country faced or will face the situation at least once. The window of opportunity was opened for Turkey in 2010, and it is thought that the advantages of the young population will be taken until 2050.

In this study, age structures of countries, OECD (35 countries), High-Income Countries (Per Capita Income > $12.236) (78 countries), EU (28 countries), Eurozone (19 countries), World (217 Countries – World Bank DataBank), will be compared to better understand at what stage of "Demographic Window of Opportunity" in Turkey. The results will be compared with unemployment rates to reveal how effectively the opportunity window is being utilized. An evaluation will be carried out according to the results and some suggestions will be made.

2 Demography

2.1 The Definition of Demography

Demography, which is consisted of two words, "demos" (people, in Greek) and "graphein" (writing, drawing, describing) (Sobotka, et al., 2005), is defined as "Population Science" by Turkish Language Association (Turkish Language Association, 2006). However, demography addresses a wide range of disciplines and is associated with many disciplines, so this definition is too narrow and inadequate. Braun and Schubert (1996) and Zsindely and Schubert (1992) defined the demography as a science that examines population size, structure and quantitative aspects of population growth (Braun & Schubert, 1996, p. 57; Zsindely & Schubert, 1992, p. 17).

Domestic resources define demography as a science that examines not only quantitative (about numbers and static) growth but also qualitative (dynamic and conditional) growth (Seyfullahoğulları, 2011, p. 3), and not only population size but also situations such as birth, death, migration and their sub-concepts of marriage, divorce, education, health, etc., which are the basic variables of population (Yılmaz and Kiracı, 2016, p. 28).

2.2 The History of Demography

It is thought that Ibn-i Haldun (1332–1406) made his first scientific work in the field of demography in his work *Mukaddime*. Haldun, in the related work, stated that the population was not only interested in the numerical size but also a science related to the social, political, military and economic conditions of the population (Başar, 2013, p. 9). However, the emergence of demography as a scientific discipline began with the work of John Graunt, who was called the founder of the demography in 1662 (Vilquin, 2000, p. 49). In this study, Graunt did not only focus on births and deaths, but also discussed issues such as population age structure, life table and fertility rates in women (Wuncsh, 2000, p. 37).

After Graunt's work, studies in the field of demography have been accelerated and many studies have been done. It is approved that the first work in the field of demography is *An Essay on the Principle of Population* by Robert Malthus. However, some scholars have pointed out Louis Chevalier's work "Les Recherces et Considerations sur la Population de la France" (France's Population Explanations and Points to be Considered) as the first work in the field of demography, because they claim that it was published 20 years before Malthus's work (Kanbolat, 1998, p. 7). Although many studies have been carried out in the field of demography in the following years, it was observed that the concept of demography is first described by the French naturalist and statistician Achille Guillard in his book *Eléments de statistique humaine ou Démographie comparée* (*Human Statistics and Demography Elements*) (Joachim & Nowotny, 2000, p. 58).

2.3 The Demographic Window of Opportunity

The concepts of "Demographic Window of Opportunity" and "Demographic Dividend" which were literally introduced by Barlow are defined as an increase in working-age population (15–64) despite of a decrease in the rate of population growth (Köksel, 2016, p. 2014). The proportional increase in the range of the 15–64 age group in the total population allows the child (0–14) and the elderly (65+) to fall in dependency ratio (Mumcu & Çağlar, 2006, p. 1).

In Turkey, which enters the final stage of process of demographic transition starting from 2000s, it is expected that the growth rate of 15–64 age group will continue to grow, albeit at a steadily decreasing pace, until 2023 and the population will continue to increase in a declining manner between 2023 and 2050. It is also expected that depopulation trend will begin after the year 2050 and so, Turkey will have the elderly population and will lose the advantages because of the end of demographic window of opportunity (Karabıyık, 2009, pp. 298–299).

2.4 Demographic Population Theories

The population which is related with almost every discipline has led to the formation of various population theories over the years. Malthus's Theory of Population stated that the population should be reduced and pointed out geometric increase of the population while the foodstuffs are increasing in the arithmetic ratio. Malthus also stated that if this situation persists, food shortages will occur in the following years and that people will starve to death (Bilgili, 2016, p. 45); New Malthusianism or Neo-Malthusianism should control the population and reduce the population (Bozkurt, 2011, p. IV); mercantilists believe that if the population needs to be increased, then the economy will develop and the wealth will increase;

the physiocrats should be looked at positively as long as population growth does not affect agricultural production (Başar, 2013, pp. 11–12); classicists argue that overpopulation should be reduced because it will lead to unemployment and therefore poverty; according to the optimal population theory, population adjustment should be done according to sources (Murat, 2006, p. 36); according to the theory of demographic transition, they emphasized the necessity of establishing a relationship between different rates of birth and death and the rate of population growth (Başar, 2013, p. 19). These theories have a dynamic structure as well as the conditions of the region and the period, as well as for the developed and developing countries.

3 The Comparison of Data

Addressing only the numerical size of the population and the rate of population growth can prevent healthy results and the implementation of effective policies. If healthy results are desired, changes in age structure should be closely monitored. Changes in age structure affect the economy in two aspects. The first is the impact of the working age population on the economy, and the second is the determination of the direction of workforce supply (Gülsoy & Tekeli, 2015, s. 56).

The age structure which forms the basis of Demographic Window of Opportunity will be analyzed in order to determine at what stage of "Demographic Window of Opportunity" in Turkey.

- OECD (35 countries), High-Income Countries (Per Capita Income > $12.236) (78 countries), EU (28 countries), Eurozone (19 countries), World (217 Countries - World Bank DataBank)

The data for 2012–2016 will be compared. The results will be compared with unemployment rates, and it will be examined in order to determine at what stage of "Demographic Window of Opportunity" in Turkey and how effectively advantages of young population are being utilized.

The main age group of the "Demographic Window of Opportunity" is 15–64 age group, and source of this age group is consisted of 0–14 age group. In this case, if 0–14 age group is high, demographic window of opportunity will close later than expected, and it shows that Turkey is the second country that will be close later in terms of selected variables.

The high percentage of 15–64 age group, referred to as the young and active age group, indicates that the demographic window of opportunity will close later and that the advantages of this age group can be further exploited. However, 15–64 age group will be the elderly population of Turkey in the future and it can be thought that Turkey will face the problems by the elderly population. However, 0–14 age group is higher as shown in Tab. 1, and it will take place later in contrast with other countries; so Turkey will be able to benefit from the advantages of "Demographic Window of Opportunity" longer.

According to data for 2016, 195 people per 1, 000 people in the European Union, 202 people per 1, 000 people in the Eurozone, 174 people per 1, 000 people in high-income countries, 165 people per 1, 000 people in the OECD countries are elderly people, and on the other hand 85 people per 1, 000 people in the world, 80 people per 1, 000 people in Turkey are elderly people. In fact, 15–20 % of population are elderly people in other variables, except world average, while %4 of population are elderly people in Turkey.

Although Turkey is seen as slowly ageing country in contrast with selected variables, it is estimated that it will completely lose this advantage by the year 2050. It is also observed that today, 80 people per 1, 000 people are elderly people, this ratio will increase to 200 by the year 2050, and it will reach the level of advanced countries as a result of the projections carried out. Long-term plans must already be implemented to prevent the disadvantages of the elderly population.

When the unemployment rates, except Turkey, are compared with 2012 and 2016 it is observed that unemployment rate decreased by 1.9 % in European Union countries, decreased by 1.3 % in the Euro Area, decreased by 1.8 % in high-income countries, decreased by 1.6 % in the OECD member countries and decreased by 0.1 % in the world. However, unemployment rate in Turkey increased by 2.7 % and raised from 10.8 % in 2012; so it was the worst unemployment rate in the selected variables.

4 Conclusion

It is excepted that Demographic Window of Opportunity that Turkey has entered since 2010, will end up in 2050. It shows that Turkey has not much time to take advantage of window of opportunity, and it means that Turkey must choose the most appropriate policies to take advantage of this window. In this study, it is aimed to analyze Turkey's demographic window of opportunity is at which stage compared to other variables, and employment policy between the years 2012–2016. According to the results,

With the table of age structure, it can be observed that Turkey is the most advantageous country in each age group. This situation is an indication that the "Demographic Opportunity Window" will be closed at the latest, and Turkey can be benefit from the advantages longer. However, when unemployment rate is examined, the rates in other variables, except Turkey, declined in 2012. Therefore,

Tab. 1: The Comparison of Data. Source: WorldBank (2018), DataBank.

Range	Country	2012	2013	2014	2015	2016
Population ages 0–14 (%of total)	Turkey	26,4	26,1	25,9	25,6	25,3
	European Union	15,6	15,5	15,5	15,5	15,4
	Eurozone	15,3	15,3	15,2	15,2	15,1
	High-Income	17,1	17,0	16,8	16,7	16,6
	OECD Countries	18,6	18,4	18,3	18,2	18,0
	World	26,5	26,4	26,3	26,2	26,1
Population ages 15–64 (%of total)	Turkey	66,2	66,4	66,5	66,6	66,7
	European Union	66,3	66,0	65,7	65,4	65,1
	Eurozone	65,8	65,5	65,2	64,9	64,7
	High-Income	67,0	66,7	66,5	66,2	66,0
	OECD Countries	66,3	66,1	65,9	65,7	65,5
	World	65,6	65,6	65,6	65,5	65,5
Population ages 65 and above (%of total)	Turkey	7,4	7,5	7,6	7,8	8,0
	European Union	18,1	18,4	18,8	19,1	19,5
	Eurozone	18,9	19,2	19,6	19,9	20,2
	High-Income	15,9	16,3	16,7	17,0	17,4
	OECD Countries	15,2	15,5	15,8	16,2	16,5
	World	7,8	8,0	8,1	8,3	8,5
Unemployment rate %	Turkey	8,1	8,7	9,9	10,2	10,8
	European Union	10,4	10,8	10,2	9,4	8,5
	Eurozone	11,3	11,9	11,6	10,8	10,0
	High-Income	8,0	7,9	7,3	6,6	6,2
	OECD Countries	7,9	7,9	7,3	6,8	6,3
	World	5,6	5,6	5,4	5,5	5,5

the number of unemployed people increased every year depending on the increasing rate in unemployment in Turkey. It shows Turkey's most advantageous position, but they could not benefit fully from the advantages of Demographic Window of Opportunity. Turkey's demographic window of opportunity will close in 2050, and it is considered to be the most advantageous situation according to selected variables. The most effective ways to benefit from the Demographic Window of Opportunity are:

- Increase in birth rates should be ensured by investigating the causes of the decrease in increase rate in 0–14 age group. This will allow the demographic window of opportunity to stay open longer.

- 0–14 age group is now consumer, but also producer of the future. Therefore, investments made for this age group can be regarded as investments made for the future.
- It is seen in the unemployment rates that the 15–64 age group is not effectively benefit. Turkey should follow policies to increase employment opportunities to benefit from the advantages of young population. Thus, it can be most effectively utilized from the opportunity window if supported by qualified workforce.
- If the ration of population ages 65 and above is low, the resources to be allocated to the elderly population are education, health, employment and so on, allowing more resources to be provided to these fields. On the other hand, when Turkey's ageing process is already considered, long-term policies should be followed for the elderly people.
- Women should be given vocational training and employment should be provided on the labour market.
- The educational opportunities of the young population should be increased and legal arrangements should be made to allow employment in a market that produces high value-added products.
- In parallel with increasing educational attainment, high-value added high-tech products can be produced and these products can be exported if R&D expenditures and GDP increases.

Turkey has a chance to follow more effective policy towards Demographic Window of Opportunity, which will not last forever, because Turkey experienced this situation later than Europe and observed the problems that they face. If reasonable policies are not followed, the "Demographic Opportunity Window" can be terminated with "Demographic Disaster".

References

Başar, E. (2013). *Demografiye Giriş* (2.Baskı b.). Ankara: Gazi Kitapevi.

Bilgili, Y,. (2014). *Karşılaştırmalı İktisat Okulları, Makro İktisadın Esasları (14. Baskı)*. İstanbul: 4T Yayınları.

Bozkurt, Ö. K. (2011). Uluslararsı Nüfus ve Kalkınma Konferansı(ICPD, 1994) Eylem Programının Türkiye'de Uygulanan Sağlık Politikalarına Yansmalarının Toplumsal Cinsiyet Perspektifinden İncelenmesi. Ankara: Kadının Statüsü Genel Müdürlüğü.

Braun, T., & Schubert, A. (1996). Power Position In Science Journal- Their Gatekeeping, Demography, Ecology and Accessibility. *Les sciences au Sud : état des lieux*, 6, 51–64.

Gülsoy, G., & Tekeli, S. (2015). Nüfusun Yaşlanması ve Ekonomik Büyüme İlişkisi: Türkiye Üzerine Bir Analiz. *Amme İdaresi Dergisi, 48*(1), 35-87.

Joachim, H., & Nowotny, H. (2000). Demography And Sociology. *Position of Demography Among Other Disciplines*, 73-80.

Kanbolat, Y. (1998). *İktisaden Geri Kalmış Ülkelerde Nüfus Sorunu*. Ankara: Güldikeni Yayınları.

Karabıyık, İ. (2009). Avantaj ve Dezavantajlarıyla Genç İşsizliğinin Değerlendirlmesi. *Erzincan Üniversitesi Hukuk Fakültesi Dergisi*, 298-299.

Köksel, B. (2016). Demografik Fırsat Penceresi'nden Türkiye'de İstihdam ve İşsizlik. *Uluslararası Sosyal Araştırmalar Dergisi, 9*(43), 2013-2022.

Mumcu, O.tr, & Çağlar, E. (2006). *Türkiye'nin Nüfusu Zenginlik Kaynağı Olabilir mi?* TEPAV.

Murat, S. (2006). *Dünden Bugüne İstanbul'un Nüfus ve Demografik Yapısı*. İstanbul: İstanbul Ticaret Odası. Publicatin no:2006-49.

Seyfullahoğulları, Ç. A. (2011). *Türkiye'nin İller Bazında İktisadi ve Demografik Farklılıkları (Kovaryans Analizi Yaklaşımı)*. İstanbul: Beta Basım Yayım.

Sobotka, T., Winkler-Dworak, M., Testa, M. R., Lutz, W., Philipo, D., Engelhardt, H., & Gisser, R. (2005). Monthly Estimates of the Quantum of Fertility: Towards a Fertility Monitoring System in Austria. *Vienna Yearbook of Population Research, 3*, 109-141.

The World Bank. (2018). Databank. On 03 14, 2017 World Development Indicators: RecivedFromhttp://databank.worldbank.org/data/reports.aspx?Code=NY.GDP.MKTP.KD.ZG&id=1ff4a498&report_name=Popular-Indicators&populartype=series&ispopular=y#

Turkish Language Association. (2006, September 26). Güncel Türkçe Sözlük. On April 2017, 25 Received From TDK: http://www.tdk.gov.tr/index.php?option=com_gts&arama=gts&guid= TDK.GTS.5b07f0d12eb4d0.60239159 Vilquin, E. (2000). History Of Demography. *Position of Demography Among Other Disciplines*, 49-60.

Wuncsh, G. (2000). Demography: A Discipline Somewhere Between Philosophy And Social Care. *Position of Demography Among Other Disciplines 37*, 37-41.

Yılmaz, H., & Kiracı, A. (2016). *Siirt İlinin Demografik Yapısının Türkiye İle Karşılaştırılması ve Bölgesel Kalkınma Açısından İncelenmesi*. İstanbul: Kutlu Yayınevi.

Zsindely, S., & Schubert, A. (1992). The Demography of Journals. *New Library World, 93*(4), 17-21.

Fatih Çağatay Cengiz

The Politics of Development of Turkey's Indigenous and National (*Yerli ve Milli*) Defense Industry

1 Introduction

According to the 2017–2021 Strategic Plan prepared by the Undersecretariat for Defense Industries, "global activity and technological depth" is the fundamental vision underlying Turkey's current defense strategy. The 2012–2016 Strategic Plan, however, had only specified "technological supremacy" as its underlying goal. The most recent Strategic Plan outlines four parameters for achieving the aforementioned vision: "production and service efficiency," "abolishment of uncertainties in order to create a competitive environment," "creation of an environment where defense and security industry would interact with all sectors," and "determination of a competitive economic environment that has technological development and depth" (Savunma Sanayi Müsteşarlığı, 2017: 77). This emphasis on "technological depth" parallels Turkey's recent discourse on developing an "indigenous and national defense industry." The production of MİLGEM (*Milli Gemi*, National Ship), the ALTAY main battle tanks, T129 ATAK attack helicopters, Anka and Bayraktar unmanned aerial vehicles, new type patrol boats, fast response boats, and national infantry rifles since 2004 shows Turkey's aspirations to become self-sufficient defense-wise (Savunma Sanayi Müsteşarlığı, 2017: 5). However, Turkey's dependence on NATO allies for the import of high-tech defense products, especially in the aircraft and naval industry, is an impediment to the realization of its self-sufficiency. Despite growing tensions between the USA and Turkey around Syria, it seems unlikely that Turkey will delink from NATO anytime soon. With this in mind, this chapter sheds light on the dependent development of Turkey's defense industry from the early Republic onwards. It is suggested that, even though the development of Turkey's defense industry highlights a dependent relationship, the level of this development contributes to Turkey's aspiration to be a regional power.

2 Literature on Turkey's Aspiration to be a Regional Power

There is a plethora of literature on Turkey's attempts to fulfill its role as a regional power in the post-Cold War period. According to Dinçer and Kutlay, a regional power "should have a 'role definition' in parallel with this intention, and it should make it felt in its relations with other actors of the region" (2012: 10). In other words, material power resources and military power should be complemented by economic, diplomatic, and institutional capacity. In addition, regional competitors and global actors should accept the country as a regional power. Lastly, the foreign policy setting and implementation should produce the expected results (ibid).

In this regard, Şaban Kardaş locates Turkey as a regional power by arguing that it possesses overwhelming material capabilities compared to its neighbors in terms of population, size, gross domestic product, and military spending; exerts influence to resolve regional conflicts such as in Palestine-Israel and Syria; and presents a model for other Middle Eastern countries to follow and potential leadership (2013: 648–651). Emel Parlar Dal (2016) agrees with Kardaş (2013) regarding Turkey's increasing activity in the region and tests it with Daniel Flemes's framework on regional power status using four preconditions similar to Dinçer's and Kutlay's analysis (2012): "(1) claim to leadership; (2) possession of necessary power resources (material and ideational); (3) employment of material, institutional and discursive foreign policy instruments; and (4) acceptance of leadership by third parties" (2016: 1426). Dal concludes that Turkey can be considered an "emerging regional power," but also has some reservations to this argument, suggesting that Turkey's position as a regional power has declined in the context of regional and global crises (2016: 1445).

In addition, Dinçer and Kutlay (2012: 37) highlight a discrepancy between Turkey's "non-material power elements" such as "capability of controlling information structures, agenda-setting power, role perception, and regional acceptance." Similarly, Ziya Öniş and Mustafa Kutlay warn that Turkey's "benign regional power" is restricted by its retreat from democratization and its structural economic problems, such as its current account deficit and energy dependence (2013: 1415–1420). Öniş suggests in another article that the ruling party's "insufficient knowledge and expertise about the domestic politics of Arab states" has dragged Turkey into over-engagement with sectarian conflicts in the region, siding with Sunni Muslims such as the Muslim Brotherhood in Egypt and the Sunni opposition against Basher al-Assad in Syria (2014: 212–216). Turkey's 2008 support for Iran's nuclear program, its violent confrontation with Israel regarding the Gaza flotilla in 2010, and its rapprochement with Sudanese President Omar al-Bashir, who was charged with genocide in Darfur by the international criminal court, led Svante Cornell to share the views of the Justice and Development Party's critics, who argue that Turkish foreign policy is at an "axis shift" (2012: 13–16). Conversely, however, Taha Özkan does not identify Turkey's over-engagement with the Middle East as a foreign policy "axis shift" (2010). Rather, Özkan suggests that Turkey's activity in the Middle East is an attempt to root out "Western-centered and Eurocentrist conditionedness" by adapting itself to the *zeitgeist* of international politics. This policy is the result of Turkey replacing US diplomacy in the region following 9/11 and the political cost of US unilateralism in Iraq, Afghanistan, Gaza, and Lebanon (2010: 18).

It is undeniable that asserting oneself as a regional power necessitates a balanced combination of soft and hard power. It can be argued that, at least until 2011, Turkey could exert its power in the region by utilizing its material and non-material capabilities. These capabilities include: diplomatic power, in which the number of career personnel increased from 882 in 2000 to 1146 in 2011 and the number of embassies increased from 91 in 2000 to 114 in 2011; economic power, in which $135 billion worth of products were exported in 2011; and soft power, in which Turkish soap operas and state-run TRT Arabic have increased interest in Turkey (Dinçer and Kutlay, 2012).

However, since the 2011 eruption and expansion of the Arab uprisings, Turkey has relied heavily on the development of its hard power and military capabilities. The government's recent discourse on the need to develop an indigenous national defense industry marks Turkey's attempt to become militarily self-sufficient. The development of its defense industry is also an element of Turkey's foreign policy, which aims for regional expansion through exporting arms. Having said so, however, Turkey has had a historical desire for self-sufficiency that goes back to the era of the early Republic, when local defense industry production was initiated. The following section looks at this initial stage before moving to the underdevelopment of the Cold War period. Following on, the next section then analyzes how Turkey's 1980s transition to an export-led economy impacted the development of the defense industry and its evolution in the 1990s. The final section addresses the Justice and Development Party's attempts to develop an indigenous national defense industry and Turkey's increased military spending after 2008. In conclusion, this chapter outlines how the structural problems of Turkey's economy and disagreements with NATO allies have led to a discrepancy between what Turkey can do and achieve with respect to its defense industry.

3 The Initial Stage of Establishing a Defense Industry in Republican Turkey

During the republican era, a retrospective nationalist narrative posited that the lack of a national defense industry in the Ottoman era and the early period of the Republic caused Turkey to have a dependent relationship with Axis Powers and the Soviet Union. Therefore, the state elite, especially Kazım Özalp and Recep Peker, the National Defense Ministers, highlighted the necessity to develop a national defense industry (Kurç, 2013: 120-121). For instance, Kazım Özalp argued that purchasing naval ships from foreign suppliers would be of no use if the country could not develop its own naval industry. On the other hand, Recep Peker stressed the need to integrate other industrial sectors, such as steel and copper production, into the defense industry (ibid: 121). Thus, a "mixed approach" was adopted in which the national bourgeoisie and foreign capital worked together to develop Turkey's defense industry.

With the help of the German capital of *Junkers*, Turkey established its first aircraft factory in Kayseri in 1926 (ibid: 122). The Turkish government then declared that it would grant a monopoly to the German company in terms of supplying all aircraft orders (Dervişoğlu, 2014: 71). Consequently, the state's organic relations with foreign capital contributed to the development of a local, *if not* national, defense industry particularly in the central Anatolian cities of Ankara and Kırıkkale. According to Mehmet Evsile, who documented all defense-related factories' set up between 1920 and 1938, the German Fritz Werner company helped establish a rifle factory in Kırıkkale in 1935; the German Guttehoffmangshütte-Raynmetal Borsing companies helped set up an artillery factory in Kırıkkale in 1937; the French Melinit company helped open a pistol cartridge factory in Silahtarağa in 1934; the German Nielsen Winther company helped establish an artillery ammunition factory in Kırıkkale in 1925; the French Melinit company helped open a gunpowder and explosives factory in Elmadağ; and the German Köln-Ruttwell company helped set up a gunpowder factory in Kırıkkale in 1936 (1992: 81-144). In addition, the German Gutte HoffnungsHutte-Demag company and the Swedish Landes Krona company helped establish rolling plant (*haddehane*) brass-casting companies in Kırıkkale in 1926 (ibid: 152-164).

Until 1948, however, Turkish soldiers' military education was based on the Prussian military model, which promoted the idea of *Millet-i Müsellaha* (The Nation in Arms) and the integrity of military education, munitions (*levazım*), and fulfillment (*ikmal*), which affected the development of the Turkish defense industry under German influence (Özcan, 2010: 197-201). In 1924, a German-supported Dutch firm – Ingenieurskantoor voor Scheepsbouw – was awarded one of the first submarine bids (ibid: 197). In 1929, Turkey commissioned Italian shipyards to provide two destroyers, two submarines, and three submarine chasers for the Turkish Navy (Güvenç and Barlas, 2010: 237). While French military experts trained Turkish soldiers on defensive war, fortress tactics, and line of defense, German military experts trained them in offensive war tactics and command and control procedures. Until 1948, the Turkish military adopted German technical manuals (*talimname*) (Özcan, 2010: 198-201). Furthermore, when the

Air Force Academy was established in 1937, it was under the influence of the UK (Güvenç, 2010: 260).

Apart from foreign capital, local bourgeoisie also participated in the development of the defense industry. Save Vecihi Hürkuş, a military pilot who unsuccessfully attempted to obtain a license for his "Vecihi XIV" aircraft, the state supported local bourgeoisie. While Şakir Zümre used his patronage relations with Mustafa Kemal to establish a munitions factory in İstanbul in 1924, Nuri Demirağ established an aircraft factory that produced NuD 36 planes in the 1930s (Kurç, 2013: 124). Nuri Demirağ also led the National Development Party, Turkey's first opposition party in the multi-party period, which was established in 1945.

Additionally, the early republican period also saw attempts to mobilize the people to assist in the development of the defense industry. When the Turkish Aeronautical Association (*Türk Tayyare Cemiyeti*) was established in 1925, the state mobilized a public campaign to encourage donations to the association (Aydın, 2011: 65–66). Thus, Turkey's initial attempt at developing a national defense industry was shaped by a policy of cooperation between the state, foreign capital, and local bourgeoisie. As argued, this was the result of a nationalist rhetoric that aimed to control national economic development. Nevertheless, the early Republic relied on the collaboration between local and foreign capital to develop its defense industry. Later on, the politics of defense industry development was shaped to a large extent by an international context in which Turkey's geopolitical position played a key role in rapprochement with the USA.

4 Modernization of Defense Industry in the Context of the Cold War Period: A Dependent Relation with the USA

In the context of Soviet expansion in the Middle East and the US containment policy, the Cold War period saw an "alliance" form between the USA and Turkey (Güney, 2005: 341–342). While the 1947 Truman Doctrine incorporated Greece and Turkey into US geopolitical interests in the Cold War period, the Marshall Plan bestowed US economic assistance on Turkey (ibid: 342). Nevertheless, Turkey's entry into NATO in 1952 completely transformed the track of the country's defense industry development. Turkey joining this alliance led to the demolition of its local defense and aviation industries, as Turkey was limited in producing, developing, and maintaining defense equipment on its own (Mevlütoğlu, 2016: 10). Rather, Turkey chose to gain US security assurance, which went hand in hand with arms dependency (Bağcı and Kurç, 2017: 42). As a NATO member, its infrastructural development was tied to US interests. In other words, political and bureaucratic elites were unable to make long-term investments in the defense industry, thus fortifying US hegemony in the Turkish defense industry. A Turkish General plainly expressed the Turkish army's dependence on US armament in the Cold War period:

> Every year we used to submit a list of our needs to the USA. These lists were unnecessarily long, covering everything from helmets to batteries, from ropes to tanks or anti-aircrafts. The rule was to ask for as much as possible. The main reason was that we had no armament policy of our own, nor any national objectives, nor even any idea of what we really needed. The Americans shipped over whatever they thought necessary and, regardless of their use, we were too pleased to be at the receiving end. What's more, everything was donated. (…) For instance, the M-48 tanks that were replaced by M-60 in the US Army were shipped to Turkey. Two thousand Reo trucks and 10,000 jeeps, even if they dated from World War I, were also welcome. (…) We had so little planning that we had to be reminded by the Americans which part in the warships or aircraft to replace and when. All the details were recorded in their computers which alerted them, for instance, replacements had to be made on the F-100s and the warships. Sometimes we would get huge boxes, and we wouldn't know what to do with them until the replacement instructions arrived (as cited in Kurç, 2017: 262).

In the wake of the 1964 Cyprus crisis, Turkey was forced to re-develop its local defense industry as – according to the Truman Doctrine of 1947 – Turkey was forbidden from employing US equipment in a military operation without the permission and approval of the USA. Therefore, it was only in the 1970s that the Turkish state began supporting the re-development of its defense industry. In 1970, the Air Force Support Foundation was established; in 1972, the Naval Force Support Foundation was established; and in 1974, the Land Force Support Foundation was established (Mevlütoğlu, 2016: 11).

In 1987, the law establishing the "Foundation for Strengthening the Turkish Armed Forces" (*Türk Silahlı Kuvvetlerini Güçlendirme Vakfı*, TSKGV) granted the Foundation exemption from "[c]orporate tax (except for its economic enterprises), [i]nheritance and transfer taxes concerning donations and assistance it receives, [s]tamp tax concerning all of its transactions" (Parla, 1998: 44). According to Taha Parla, TSKGV together with OYAK – the pension fund of the Turkish Armed Forces – played a key role in militarizing the Turkish economy by integrating military capital with private capital (ibid: 49). The organic fusion produced a new mode of state-protected capital accumulation, which led to a reduction in costs and monopoly benefits (ibid). The Foundation's subsidiaries – such as ASELSAN, TAI, ROKETSAN, and HAVELSAN, which are currently some of the biggest companies in Turkey (See Section 5) – institutionalized this capital accumulation.

In the neo-liberal era, the institutionalization of military capital accelerated in a Turkey under the 1980s military rule. The establishment of the Undersecretariat of Defense Industries in 1985 and the creation of a new source of state funding in the context of a neo-liberal economic transition illustrated a negotiated consensus between the civilian rule and the military. That is to say, while the military rule only officially lasted from 1980–1983 and a civilian government came to power under Turgut Özal after, the military continued to hold power through "exit guarantees" that offered influence in decision-making processes around armaments and exemption from parliamentary control over the budget (Réal-Pinto, 2017). The Undersecretariat participated in joint ventures with American companies in the 1980s and 1990s for key military projects, such as "the armoured infantry fighting vehicle, F-16 electronic warfare, the HF-SSB Radio Communications system, basic trainer aircraft, light transport aircraft, helicopter and multiple launch rocket system" (Sezgin, 1997: 391). The period also witnessed Turkish arms production companies' entry to the market, a process in line with Turkey's export-led industrialization. Nevertheless, the politics of defense industry development was still shaped by joint ventures with mainly US firms, thus creating a dependent relationship. The Cold War period fortified dependence between Turkey and the USA in both politics and security matters.

5 Turkey's Defense Industry Adaption to the post-Cold War Period and Its Development under the AKP

Between 1985 and 1990, Turkey's procurement model was based heavily on direct foreign purchase. From the 1990s onward however, the co-production and local development model began to dominate defense industry procurement (Bağcı and Kurç, 2017: 44). It was a qualitative change from "the import of completed major platforms and spare parts to … critical technologies at sub-system and component levels that are used in indigenous platforms" (Bağcı and Kurç, 2017: 45). It can be argued that this qualitative change in the development of the defense industry was instrumentalized by state elites to fortify Turkey's role as a regional power, as Turkey's soft power apparatus was complemented by the development of military power. As Rafal Wiśniewski points out, the development of Turkey's defense industry has been key to AKP foreign policy, a policy that seeks to diversify international partners without completely breaking from NATO (2015: 217).

In 1997, Turkey's military expenditure as percentage of gross domestic product (GDP) stood at 4.1, the highest ratio for military expenditure between 1988 and 2016. It can be argued that low-intensity warfare with the PKK increased Turkey's military expenditure in the 1990s. Nevertheless, it can also be suggested that

Tab. 1: Fifteen Biggest Countries in Military Spending in 2017. Source: SIPRI Military Expenditure Database, 2018a.

Country	Military Spending in 2017 (current bn., $)	Military Spending as of GDP (%)
USA	610	3.1
China	228	1.9
Saudi Arabia	69	10
Russia	66	4.3
India	64	2.5
France	58	2.3
UK	47	1.8
Japan	45	0.9
Germany	44	1.2
South Korea	39	2.6
Brazil	29	1.4
Italy	29	1.5
Australia	27	2.0
Canada	21	1.3
Turkey	18	2.2

the democratization process triggered by potential EU ascension contributed to decreasing military expenditure as a share of GDP, rather than actual spending on military. Since then, Turkey has shown a downward trend in military expenditure as percentage of GDP, standing at only 2.2 % in 2017 according to SIPRI (See Tab. 1). However, the data must be investigated cautiously, as it, firstly, reflects only the official budget rather than actual military expenditure (2018a). Taking into account Turkey's military expenditure in the context of the military operation in northern Syria/Afrin, which commenced in January 2018, these numbers could be higher. Secondly, Turkey's military expenditure increased by 46 % between 2008 and 2017, thus placing the country second only to China in increased military expenditure among the top 15 military spenders (SIPRI, 2018a). Thirdly, Turkey has been one of the biggest military spenders in the world. While the USA is the biggest military spender overall – with $610 billion in 2017 – Turkey spent $18 billion in 2017.

Moreover, Turkey is one of the largest arms importers in the world, ranking 10th with an expenditure of $13.1 billion between 2000 and 2017. Globally, India ranked first among arms importers, spending $46.8 billion between 2000 and 2017, followed by 'rising star' China with $35 billion. It is also interesting

The Politics of Development – Turkey's Defense Industry   107

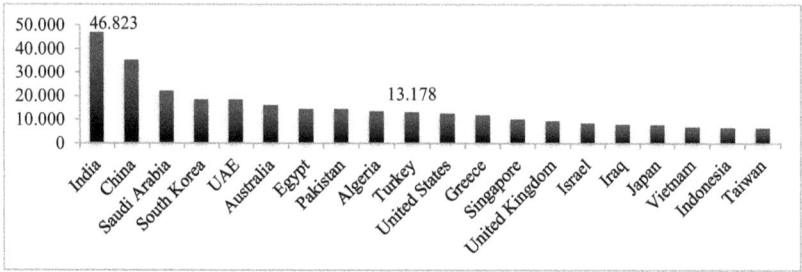

Fig. 1: Twenty Biggest Arms Importers Between 2000 and 2017 (mn., $). SIPRI adopts a unique methodology to compare the sale of conventional weapons, known as a trend-indicator value (TIV). "The TIV is based on the known unit production costs of a core set of weapons and is intended to represent the transfer of military resources rather than the financial value of the transfer ... SIPRI TIV figures do not represent sales prices for arms transfers. They should therefore *not* be directly compared with gross domestic product (GDP), military expenditure, sales values or the financial value of export licences in an attempt to measure the economic burden of arms imports or the economic benefits of exports. They are best used as the raw data for calculating trends in international arms transfers over periods of time, global percentages for suppliers and recipients, and percentages for the volume of transfers to or from particular states" (SIPRI, 2018b). The figures used in this paper are given only for *comparison*. The unit of sales and purchases (million dollars) represents TIV calculations. Source: SIPRI Arms Transfers Database, 2018b.

to note that the United Arab Emirates – a tiny country with a population of less than 10 million in 2016 – imported more arms than countries such as Pakistan, Turkey, the USA, the UK, Israel, Japan, and Indonesia (Fig. 1). Turkey has mainly imported arms – primarily aircrafts and ships for its armaments – from the USA, Germany, South Korea, Israel, Italy, and the UK (SIPRI, 2018b).

On the other hand, Fig. 2 shows that Turkish arms exports are marginal compared to the USA, Russia, and China. While Turkey exported $1.9 billion in arms between 2000 and 2017, the USA amounted $143 billion in that same period. Turkmenistan, Saudi Arabia, the UAE, Malaysia, and Pakistan were main destinations for Turkish arms exports between 2000 and 2017, which mostly consisted of armored vehicles and ships (SIPRI, 2018b). ASELSAN and Turkish Aerospace Industries (TAI) were the biggest arms-producing companies in the world, amounting $1.2 and $1.1 billion in arms sales in 2016. ASELSAN, which ranked 67th in the SIPRI Top 100 in Arms Industry Database, managed to make a $263 million in profit in 2016 (SIPRI, 2018c). The company's regional expansion follows Turkish foreign policy in the Middle East and Central Asia. The

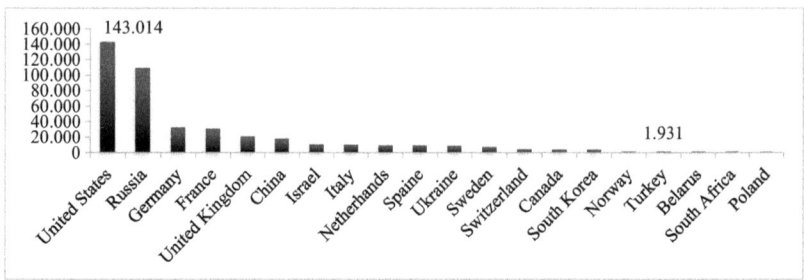

Fig. 2: Twenty Biggest Arms Exporters between 2000 and 2017 (mn., $). Source: SIPRI, Arms Transfers Database, 2018b.

company not only established a joint venture with King Abdullah II's Design and Development Bureau in Jordan in 2014, but also set up a factory in Kazakhistan in 2013 (Wiśniewski, 2015 221).

This is why Turkey has increasingly relied on the development of its local and national defense industry. According to the US-based intelligence platform Stratfor, there are three main reasons for Turkey to lean towards its local defense industry. The first relates to the economy under the AKP, where development of the local defense industry increases savings and beefs up economic growth. When Figs. 1 and 2 are compared, it is clear that Turkish arms imports are six times greater than its exports, thus creating a trade deficit. The second reason relates to Turkey's divergent foreign policy from the USA and Germany, who are key suppliers for Turkey's arms. Thus, the US Congress blocked the sale of drone technology to Turkey, and Germany rejected Turkey's request to purchase arms due to the deterioration of human rights in the country. Lastly, Turkey has interest in stimulating its defense industry, as it believes that such will bring Turkey in line with the Gulf Cooperation Council countries (Stratfor, 2017). Nevertheless, there are numerous reasons why it is not easy for Turkey to achieve autarky in defense industrialization, including: institutional inconsistency and rivalry between civilian and military institutions around procurements; Turkey's need to cooperate with international actors; and its emulation of the USA in industrial development (Kurç, 2017). While Turkey can export less sophisticated armored vehicles and artillery, high-value products such as aircraft and ships constitute nearly half of Turkish defense imports (Wiśniewski, 2017: 220). Moreover, Turkey's bid for defense autarky is used not only for showing its prowess as a regional power but also for patronizing nationalism, thus benefiting from "political capital" (Bağcı and Kurç, 2017: 40). In addition, defense industry development translates not only into a domestic political rent for the governing party, but also contributes to strengthening Turkey's hand in international negotiations and affirming nationalist feelings (Réal-Pinto, 2017). When the French Parliament recognized the 1915 Armenian genocide in 2001, Turkey froze all French military contracts; this same policy was applied to Israeli contracts in 2010, after nine Turkish citizens were killed by Israeli forces on the Mavi Marmara (ibid).

Nonetheless, defense industry development contributes to clientalist relations between the government in power and state-sponsored firms. The Nurol Company, the shipbuilding company Yonca-Onuk, and Ethem Sancak, who owns civilian and military vehicle producer BMC, have direct personal relationships with the AKP, which create "a bond of political allegiance and dependence" (Réal-Pinto, 2017). Sancak Holding also partners with the British Motor Corporation in producing around 1,000 Altay tanks in order to replace the Turkish army's old tanks (Gürcan, 2018). Nevertheless, even though Turkish defense companies will produce the sub-systems for the Altay tank project, the engines will be produced by a foreign partnership between the German Rheinmetall Group and the British Motor Corporation (ibid.). Both state-sponsored and traditional big capitalists in Turkey benefit from the politics of development of an indigenous national defense industry. Otokar – one of Koç Holding's firms – sells military equipment to the state. "Mercedes-Benz, MAN, STFA-Savronik, Alarko Holding, ... BMC (Çukurova Holding), TEMSA (Sabancı Holding), fnss (Nurol Holding), Nurol Teknoloji, VESTEL Savunma, OYTEK (OYAK Teknoloji), KALE Holding, NETAŞ, Siemens, and Yakupoğlu Deri Ticaret A.Ş." participate in defense tenders (Akça, 2010: 25). However, big capital and the government have not always been in harmony following the eruption of the Gezi protests in 2013. The Koç Group company Otokar was sidelined for a big contract in the Altay tank project due to allegations that the relations between Koç Holding and the government had soured (Gürcan, 2018). This clearly shows that defense industry development is not immune to political struggles within the state apparatus.

6 Conclusion

The development of Turkey's defense industry is bound to political calculations rather than economic rationality. In other words, it is the political context that shapes the development of the local defense industry. While the early republican period relied on a nationalist policy in developing a local defense industry, Turkey's alignment with NATO in 1952 completely transformed the understanding of this development. Turkey's defense industry development was tied to US interests in the region until the eruption of the Cyprus issue in the 1960s. Turkey's 1974 military intervention paradoxically contributed to the re-development of its defense industry in the context of a US embargo. Turkey's transition to an export-led economy also brought new local arms producers into the market. Moreover, Turkish defense companies' increasing visibility was viewed by the state elite as a political rent to materialize new foreign policy in the region. However, even though Turkey has made considerable improvements in fortifying its defense industry since the early Republic, the structure of the Turkish economy – exports depending on imports – hinders self-sufficiency. Thus, Turkey can produce indigenous defense items but these are not national products. As it is unlikely that Turkey will delink from the NATO partnership and link up with Eurasian partners such as Russia and Iran in the near future, dependent relations with the USA will persist. Turkey's pendulum between NATO and Eurasian countries will be contingent on a political process in which the military and political elites cannot easily coopt.

References

Akça, İ. (2010, July). *Military-Economic Structure in Turkey: Present Situation, Problems, and Solutions.* translated by E. İlhan, & E. Kalaycıoğlu. TESEV: İstanbul.

Aydın, A. F. (2011). "Tayyareden Uçağa: Milli Hava Sanayinin Kuruluşunda Türk Halkının Yaptığı Bağışlar" [From airplane to aircraft: donations of Turkish people in the process of establishing national aviation industry], *Karadeniz Araştırmaları*, 8(31): 51–84.

Bağcı, H. and Kurç, Ç. (2017). "Turkey's strategic choice: buy or make weapons?" *Defense Studies*, 17(1): 38–62.

Cornell, S. E. (2012). "What drives Turkish Foreign Policy?" *The Middle East Quarterly*, 18: 13–24.

Dal, E. P. (2016). "Conceptualising and testing the 'emerging regional power' of Turkey in the shifting ınternational order". *Third World Quarterly*, 37(8): 1425–1453.

Dervişoğlu, F. (2014). "İstikbalini Göklerde Arayan Ülke ve Türk Havacılık Sahasında Alman Menfaatleri Işığında Bir Ortaklık: Tomtaş [The country that searches its future in the Sky and a Partnership in the light of German Interests in Turkish Aviation Field: Tomtaş]". *Cumhuriyet International Journal of Education*, 3(3): 68–82.

Dinçer, O. B. and Kutlay, M. (2012). *Turkey's Power Capacity in the Middle East: Limits of the Possible*, International Strategic Research Organization (USAK) Reports No: 12-04. USAK Publications: Ankara.

Evsile, M. (1992). *Atatürk Devri Harp Sanayii* [War Industry in Atatürk Era], (Doctoral Dissertation). The Graduate School of Social Sciences of Fırat University: Elazığ.

Güney, A. (2005). "An anatomy of the transformation of the US-Turkish alliance: from "Cold War" to war on Iraq". *Turkish Studies*, 6(3): 341–359.

Gürcan, M. (2018). "Turkey still working to get Altay tanks rolling". *Al-Monitor*, https://www.al-monitor.com/pulse/originals/2018/04/turkey-army-will-use-most-expensive-tank.html, (12.04.2018).

Güvenç, S. (2010). "ABD Askeri Yardımı ve Türk Ordusunun Dönüşümü: 1942–1960" [US Aid and the Transformation of the Turkish Military: 1942–1960], in E. B. Paker and İ. Akça (eds.), *Türkiye'de Ordu, Devlet ve Güvenlik Siyaseti* [The Military, the State, and the Security Politics in Turkey], pp. 255–284, İstanbul Bilgi Üniversitesi Yayınları: İstanbul.

Güvenç, S. and Barlas D. (2010). "Bir Cumhuriyet Kurumu Yaratmak: Atatürk'ün Donanması, 1923–1939" [Creating a Rebublican Institution: Atatürk's Navy, 1923–1939], in E. B. Paker and İ. Akça (eds.), *Türkiye'de Ordu, Devlet ve Güvenlik Siyaseti* [The Military, the State, and the Security Politics in Turkey], pp. 223–253, İstanbul Bilgi Üniversitesi Yayınları: İstanbul.

Kardaş, Ş. (2013). "Turkey: a regional power facing a changing international system". *Turkish Studies*, 14(2): 637–660.

Kurç, Ç. (2013). *Critical Approach to Turkey's Defense Procurement Behavior: 1923–2013* (Unpublished doctoral dissertation). The Graduate School of Social Sciences of Middle East Technical University: Ankara.

Kurç, Ç. (2017). "Between defence autarky and dependency: the dynamics of Turkish defence industrialization". *Defense Studies*, 17(3): 260–281.

Mevlütoğlu, A. (2016). *Türkiye'nin Savunma Reformu: Tespitler ve Öneriler* [Turkey's Defense Reform: Findings and Suggestions], No: 164, SETA: Ankara.

Öniş, Z. (2014). "Turkey and the Arab Revolutions: boundaries of regional power influence in a turbulent Middle East". *Mediterranean Politics*, 19(2), 203–219.

Öniş, Z. and Kutlay, M. (2013). "Rising powers in a changing global order: the political economy of Turkey in the age of BRICS". *Third World Quarterly*, 34(8): 1409–1426.

Özcan, G. (2010) "Türkiye'de Cumhuriyet Dönemi Ordusunda Prusya Etkisi" [The Prussian Effect on the Military of Republican Era in Turkey], in E. B. Paker and İ. Akça (eds.), *Türkiye'de Ordu, Devlet ve Güvenlik Siyaseti* [The Military, the State, and the Security Politics in Turkey], pp. 176–221, İstanbul Bilgi Üniversitesi Yayınları: İstanbul.

Özkan, T. (2010). "Turkey, Israel and the US in the wake of the Gaza Flotilla Crisis". *Insight Turkey*, 12(3): 7–18.

Parla, T. (1998). "Mercantile militarism in Turkey, 1960–1998". *New Perspectives on Turkey*, 19: 29–52.

Réal-Pinto, A. G. C. (2017). "A neo-liberal exception? The defence industry 'Turkification' project", *International Development Policy Revue internationale de politique de développement*, 8: 299–331.

Savunma Sanayi Müsteşarlığı. (2017). *Stratejik Plan 2017–2021* [Strategic Plan 2017–2021], https://www.ssm.gov.tr/Images/Uploads/MyContents/F_20170606155720342529.pdf, (21.03.2018).

Sezgin, S. (1997). "Country survey X: Defence spending in Turkey". *Defence and Peace Economics*, 8(4): 381–409.

SIPRI. (2018a). *SIPRI Military Expenditure Database*, https://www.sipri.org/databases/milex, (14.03.2018).

SIPRI, (2018b). *SIPRI Arms Transfers Database*, https://www.sipri.org/databases/armstransfers, (14.03.2018).

SIPRI. (2018c). *SIPRI Arms Industry Database*, https://www.sipri.org/databases/armsindustry, (15.03.2018).

Stratfor. (May 26, 2017). "Turkey Builds a Military-Industrial Complex to Match Its Ambitions", https://worldview.stratfor.com, (15.03.2018).

Wiśniewski, R. (2015). "Military-industrial aspects of Turkish defence policy". *Rocznik Integracji Europejskiej*, 9: 215–228.

Ferihan Polat and Ömer Ayna

Moral Behavior Types of Prospective Public Administrators: A Case Study in the Department of Political Sciences and Public Administration at Pamukkale University

1 Introduction

Morality and politics are as old as human history and both are used to determine values, attitudes and codes of conduct in relation to individuals within society. The relationship between morality and politics has been a subject of many philosophic discussions since the era of ancient Greece. However, this relationship has orbited outside the realm of philosophy and after the second half of the 20th century has become the subject of experimental studies in the light of positivism.

The ideas revealed by the Enlightenment have taught individuals that they could rely on themselves and their minds without a need for either intuition or church authority in order to know good and evil. Morality consists of common values that hold the society together and make it possible to live together. Therefore, it constitutes one of the most controversial topics discussed by ancient Greek philosophers. The question whether morality should sway to what's right or good has been argued not only at the level of relations between individuals but also in terms of state-individual relations and attitudes of public administrators. However, making this argument at the philosophical level has not provided enough explanation in understanding individuals' moral attitudes and behaviors in daily life.

Moral development is one of the most important elements of personality development and concerns the development of an awareness of what is good and bad in an individual's socialization process. Therefore, morality is no longer just an area of philosophy, but it has begun to be a subject of theories and experimental studies of psychologists and social psychologists in particular. Analysis of the relationship between moral development, political attitudes and systems has over time revealed that it is almost impossible for individuals to develop a moral understanding independent from the social structure and moral values system inhabited by humans as social beings.

This study mentions Kohlberg's moral judgment theory, one of the cognitive moral development theories explaining moral development, and emphasizes the moral behavior typology for public administrators developed by Shelton S. Steinberg and David T. Austern. Supporting this typology with field research constitutes a remarkable aspect of the present study.

1.1 Cognitive Moral Development Theories

Scientists have introduced different opinions on moral and conscience development at different periods in history. The cognitive theories associated with moral development, cognitive levels and even intelligence suggest that the moral development of individuals proceeds in stages. The two most accepted theories on the stages of moral development were developed by Piaget and Kohlberg.

Piaget describes moral development using the effect of two sources such as mental development and social experience, and argues that morality consists of authoritarian and autonomous moral stages. The first stage of moral development is moral realism, which is called authoritarian ethics. At this stage, children experience moral realism because cognitive development has two fundamental flaws: egocentrism and realism. Children do not distinguish between other people's thoughts and their own and consider that both other people and themselves have the same thoughts. Children who cannot distinguish between subjectivity and objectivity in this stage think of their dreams as real things. Moral development at this stage has two characteristics: Firstly, as children are egocentric, they do not know that people may have different thoughts regarding morality. Therefore, they consider that there is only one type of moral judgment which is accepted by everyone. Secondly, as children are realists, they cannot differentiate social rules or psychological beliefs from physical rules. They consider moral rules as a part of nature. This kind of moral development leads to the development of a sense of justice based on authority (Güngör, 1993: 46–47).

The mind needs to be free from egocentrism and realism in order to pass into the autonomous moral stage specified by Piaget as the second stage. At this stage, children's cooperative activities with their peers are more important than their unilateral authoritarian relationship with their parents. As children grow older and the pressure of elders decreases, they have more relationships and get more opportunities to discuss many things freely with their peers. Thus, children experience a new relationship system based on cooperation rather than on command and control. Now the rules are no longer an invariant order set by God or elders, but refer to mutual agreements made between people in order to reach certain goals. The roles of children in play with their peers improve their ability to put themselves into others' shoes from time to time. When they discover that they need to have different attitudes in different roles, they realize that they have different thoughts from their friends and are in fact, different people. However, the most important thing leading to this awareness is that children develop equality-based relationships. In addition, when parents exhibit equality-based reactions to children, then children are able to put aside their ideas of moral realism, that is, considering morality as an absolutely compelling system operating outside of themselves. Children who have reached this developmental stage can understand that lying is worse than lying to parents. The sense of justice also develops at this stage, and justice is evaluated according to equality and intention (Güngör, 1993: 47–49).

1.2 Kohlberg's Theory of Moral Judgment Abilities

Kohlberg, who considered moral development to be parallel to mental thought, was a social scientist who has maintained the tradition of Piaget and developed his theory based on the cognitive learning movement. Kohlberg, who examined moral thinking in various countries including Thailand, Malaysia, Mexico and Turkey, has determined the existence of similar development in all of those countries examined. He also determined that developmental process was not dependent on a religion or lack thereof. According to Kohlberg, people in all cultures use the same basic moral concepts such as justice, equality, love, respect and authority (Onur, 2000:174).

In order to reveal the existence of six moral developmental stages, Kohlberg developed dilemmas that would determine the thought processes used by people to solve their moral problems. After he presented dilemmas that would reveal thought structures of both children and adults, Kohlberg developed three levels and six stages of moral development by grouping the options used by subjects for solving dilemmas and their reasons for selecting these options. These levels and stages are listed as follows (Mercin, 2005, p. 81):

Level 1: Pre-conventional (obedience and punishment, purely self-interest orientation)
    At this level, children are open to the concepts of good and evil, right and wrong, which are imposed by culture, and conform to the rules of a common code of conduct.

Stage 1 (4–5 years): Punishment and obedience period: At this stage, children observe what is right and wrong as a result of a specific behavior. For example, an action is perceived as morally wrong if the child who commits it gets punished or vice versa. The first stage has extremely primitive features. "The child looks for solutions to all problems with physical punishment" (Çileli, 1981: 58). The child also believes that a good behavior brings a reward. An example of this stage may be the behavior of a driver running a red light at a crossroad where there are no traffic police, or a cheating behavior of a student who realizes that the examiner will not see him/her.

Stage 2 (6–9 years): Along with some pre-conventional features, the second stage has more advanced features than the first. These features result from children's new mental abilities and role-playing skills (Çileli, 1981: 58). The understanding of an eye for an eye and a tooth for a tooth dominates at this stage. Children obey rules as long as these rules satisfy their needs. For children, everything is mutual in this period. What is "right" at this period is a concrete and reciprocal fair exchange, taking into account the needs of other people. The person in this stage has the understanding of "you get what you pay for".

Level 2: Conventional (good boy/girl, inclination to obey laws and rules)

At this level, the behaviors expected by families, groups or individuals are as valuable as a person's own beliefs. It is a matter of acceptance beyond reconciliation.

Stage 3 (10–15 years): At this stage, interpersonal compatibility or "good girl, good boy" orientation is seen as a good behavior or a behavior pleasing others, helping others and liked by others. There is stereotypical agreement with the behavior performed by the majority of people or the natural behavior. It has become important for individuals to be appreciated politely in this stage. At this stage, a good citizen should pay taxes and a good child obeys the rules set by their parents and acts accordingly.

Stage 4 (15–18 years): Individuals have an inclination to obey laws and rules. They begin to obey authority and rules and fulfill society's demands in order to protect the social system. At this stage, individuals argue that students should not cheat in exams because cheating in an exam is contrary to the rules, and although no one pays taxes, they advocate to pay taxes because paying taxes is mandatory by law.

Level 3: Post-conventional (social contradictions, universal moral principles)

This level aims to preserve moral codes, values and legal agreements with universal validity. Even if these rules contradict the group laws, it still aims to protect the rules. At this level, there is a clear effort to interpret moral values and principles.

Stage 5 (18–20 years): A good action is defined according to norms accepted by the entire community. Agreements cannot be considered "good" or "bad" unless they conflict with basic human rights such as the right to life, freedom and security. Agreements violating fundamental rights are morally invalid even if they are willingly concluded by parties. For example, human exploitation and slavery cannot be accepted even if there is an agreement between the parties. "It is necessary to protect basic human rights and freedoms such the right to life, freedom and security, even if it contradicts the majority opinion" (Çileli, 1981: 56). According to the understanding of this stage, no law can legitimize a practice that could lead to the death of a person and likewise, no one has the right to steal.

Stage 6 *(approx. 20 years)*: At the stage of universal principles, compared to social rules, personal moral values are based on abstract characteristics. At the sixth stage, the value of life is regarded as a categorical necessity, beyond all kinds of interpersonal relations. At this stage, there is an orientation towards universal ethical principles. When the value of life comes into question, it does not matter to whom it belongs, what kind of relationship, the closeness of the relationship or agreement it generates. At this stage, the following values develop:

1 Weighing your wishes by putting yourself into the shoes of all other people (including yourself) involved in a situation.
2 Then, thinking whether you would make the same decision again even if you did not know which person you would be in the situation.
3 Later on, behaving in line with the reversible decision.

Kohlberg reported that there were no judgments in line with the sixth stage among the subjects except for philosophers, but when the characteristics of the sixth stage were shown to the subjects who had developed judgment with characteristics of the fifth stage, they preferred the sixth stage (Çileli, 1981: 56).

Kohlberg argues that the moral judgment of each individual develops in the same order and periods and occurs continuously, stage by stage. Individuals generally experience the moral judgments of the developmental stage which they are in and rarely decline into the previous stages. The role of education in moral development helps individuals pass into the next stage.

This model, developed by Kohlberg, focuses on how people think rather than how they behave in the face of ethical dilemmas. Therefore, how individuals actually act when making specific decisions was not tested on the model, only the moral judgment skills were tested. The model assumes a harmony between the thoughts and actions of the individual. Although moral judgment is a necessity; honesty, altruism, and resistance to immorality are not enough for moral behaviors (Aytemiz, Seymen and Bolat, 2007: 34). As a matter of fact, studies reported a statistically significant relationship between moral judgment and moral action, but found that moral judgment was inadequate to fully explain immoral behaviors. Studies also report that although people in a higher moral stage are able to resist being oppressed for adapting to other people's judgments, there is no strong research findings supporting the assumption that individuals at a higher moral level are more honest and altruist. Therefore, the assumption that individuals at the post-conventional level can resist pressure on them for adapting to the community, in terms of moral actions, does not have strong support (Trevino, 1986: 609).

1.3 Dimensions of the Relationship between Morality and Politics

Politics is broadly defined as a struggle for power, and narrowly expressed as a subordination of the ruler over the ruled. The ability of the ruled, who has no primary relationship with the ruler, to determine his/her own destiny, transforms the relationship between the ruler and the ruled into a hierarchical structure within politics. The nature of power relations within politics – which establishes the balance in this hierarchical structure – sometimes using material force or sometimes using persuasive force does not change. Within this hierarchical structure of politics, the ruled needs to adapt to the conditions compelling his/her ego, such as obeying prohibitions and restrictions set by the ruler, accepting that some people are superior to him/her and accepting that these people have qualifications to decide about his/her own life. The ruler has to convince both him/herself and the ruled that he/she has the right to make decisions about others' lives.

There is a serious moral problem at the core of politics. Equality between people becomes a pre-acceptance if morality is thought to be in favor of people and life; however, politics, which has a hierarchical structure by nature, institutionalizes inequality between people (Ağaoğulları, 1993). Therefore, politics, which is based on the distinction between the ruler-ruled among people, is contrary to morality in principle. A variety of legitimacy theories based on a conviction that the ruler has the right to govern the ruled have been generated, and these theories are among the main topics of political science and political philosophy. The references made by legitimacy theories to neither divine nor secular sources have not been able to solve the problem of inequality within the nature of politics.

Alkan (1993) explains using moral problems at the core of politics why moral discussions play such a crucial role in political life and why moral criteria are applied more in assessing politicians and politics than in evaluating other people and events. Alkan also argues that unlike other moral problems, immorality results from the essence of politics, so has an unsolvable nature (Alkan, 1993: 109).

Another point generating a contradiction between politics and morality is politics itself is seen as a field where all interests in society intersect. One other point where politics and morality contradict each other is how politics, which assumes the function of organizing distribution relations between social classes or groups, manages this distribution conflict. Politics, which is a product of a fundamentalist contradiction between the ruler and the ruled, is the basic authority on solving contradictions in society and constitutes a social field which not only contradicts with moral values more than all other fields but at the same time needs morality more than any other field.

The third intersection point of politics and morality concerns the idea that the main objective of politics is philosophical rather than scientific and technical. Politics constitutes the most general sub-system among all sub-systems in society. As politics is responsible for keeping all the institutions of society together, ensuring their coordination and guiding them to the future, it is also responsible for the functioning of all institutions. Politics carries the burden of determining and realizing the objectives of the entire society. Therefore, adopting one objective instead of another is based entirely on moral choice. The objectives adopted or rejected as good or bad are not formulated as a result of scientific or technical necessities but through clarification of political tendencies. Although scientific and technical knowledge contribute to the implementation process after determining the objectives, we encounter each political objective as a pedigreed moral choice since the objectives themselves are propositions about the results (Alkan, 1993: 110–111).

The fourth dimension of the relationship between morality and politics is related to politics' regulatory and supervisory function over other social institutions. Emphasis on politics as a relief of moral corruption consists of another point where morality and politics interject. If there is a moral deterioration and dissolution in a society, politics is expected to solve this problem. Politics is expected to align both itself and other institutions and processes in society in terms of ethical practices, and also is considered as a starting point for actions to be taken for a return to morality (Alkan, 1993: 111–112).

The complex and multidimensional relationship between politics and morality sometimes leads to a number of wrongful and unjust judgments in the evaluation of politicians and political events. In the face of many such political events, the tendency is to place blame for immorality on politicians' personal identities. However, while there may be no contradiction between morality and politicians, there is both a contradiction and relationship between politics and morality due to the above-mentioned points and the nature of politics. Therefore, it is very important to make a distinction between whether immorality emerges in an immoral political event, due to the nature of politics, or the personality of the politician(s). In many cases, the moral problem may arise due to the specific moral nature of the politics rather than the individual morality of the politician(s). As Alkan (1993) points out, criticizing politicians by perceiving moral problems which stem from the very nature of politics as the general immorality of the politicians may cause for politics to experience an unfair loss of reputation, discourage many valuable people from going into politics and cause the quality of politicians to deteriorate (Alkan, 1993: 112).

1.4 Moral Dilemmas in Public Administration

The regulatory and supervisory function of politics on other social institutions, the fact that politics is an area where all interests in society intersect, and the institutionalization of inequality between people due to the hierarchical structure of politics by the nature of politics, has prompted many social scientists to seek answers for the questions of "what is moral in politics" and "what is moral behavior". These questions, which have been the main point of intersection between political philosophy and philosophy as a whole since ancient Greece, have only begun to be the subject of quantitative research since after the second half of the 20th century in social sciences which were born after the 19th century. This problem, firstly handled by social psychology, has, especially after 1980, become of interest to social scientists working in the field of public administration.

Applied studies conducted in the administrative area namely by Rest (1984), Trevino (1986), Ferrell and Gresham (1985), Jones (1991), Strong and Meyer (1992), Harrington (1997), Thorne (1998), Robertson and Fadil (1999), Street et al. (2001), Leonard, Cronan and Kreie (2004) have developed ethical decision-making models with reference to the theory of cognitive moral development.

1.5 Moral Behavior Typology of Sheldon S. Steinberg and David T. Austern

Apart from the ethical decision-making models mentioned above, another important study questioning moral behaviors in public administration was conducted by Sheldon S. Steinberg and David T. Austern in 1996: "Government, Ethics, and Managers: A Guide to Solving Ethical Dilemmas in the Public Sector". This study examined the reasons for immoral behaviors and dilemmas of public officials, selecting more than one thousand public officials and managers and asking about 14 cases of "moral dilemma". According to their answers, the researchers have classified three types of government practitioners: "the corrupter", "the functionary" and "the ethicist" (Steinberg and Austern, 1996: X). They have completed their study by explaining the following four conditions which they think are necessary for preventing or correcting immoral practices: training, financial audit of management, investigation and management control (Steinberg and Austern, 1996: 141).

The characteristics of the three types of ethical behaviors explained by Steinberg and Austern (1996) that constitute the subject of this study are explained below.

2 The Corrupter (Conduct Based on No Ethics)

This practitioner, unaffected by any moral or spiritual connection, will benefit from every clear opportunity for his/her own personal gain at every occasion or when sufficient incentive is provided because he/she is unprincipled (Steinberg and Austern, 1996: 76). Some of the characteristics of this person are as follows:

- Conducts immoral or illegal practices. Hides information by lying and destroys documents.
- Uses his/her chair at the public office as a tool to increase personal gain.
- Benefits from other people using their moral weakness.
- Enacts fraudulent practices against the law, hiding false and unjustified works, and using force for his/her own power (Steinberg and Austern, 1996:79).

3 The Functionary (Value-Neutral or Relativistic Conduct)

Even if something is obviously wrong, the functionary does not object or oppose to authority, acts detachedly, does not break the order and thus adapts to changes. Even if there is a moral problem, this person follows and obeys orders (Steinberg and Austern, 1996: 78). Some of his/her personality traits are as follows:

- Does not partake in illegal or immoral activities, and does not make others get involved in such activities, but also does not give inside information to his/her managers when necessary.
- Takes the principle of "I do not want any problems" and does not do any more or less than the expected, only does what is expected.
- Gives the least of the information demanded, and hides information even if it is not appropriate in order to protect the organization.
- For him/her, the job refers to "the thing that he/she devotes time until his/her retirement".
- Follows the leader even in case of personal discomfort occurring as a result of a moral problem (Steinberg and Austern, 1996: 80, 81).

4 The Ethicist (Conduct Based on Ethics That Promote the Public Good and Public Interest)

The ethicist prioritizes protection of the public interest, and does not treat people the way he/she does not want to be treated as defined by clichéd morality. Some of the concepts that he/she emphasizes are individual freedom, honesty, objectivity and equality before the law (Steinberg and Austern, 1996: 78). The ethicist has the following behavioral characteristics:

- Knows to say "no" when he/she is asked to involve in an unethical or illegal activity.
- Does not tolerate others' immoral behaviors and does not canalize them to illegal activities.

- Tries to do his/her job as best as possible and encourages others to do the same.
- Is aware of the difference between candidness and unfaithfulness.
- Is honest and open in communication, hides information solely because of legal or ethical necessity (Steinberg and Austern, 1996: 81, 82).

5 Moral Behavioral Typology of Sheldon S. Steinberg and David T. Austern: The Case from Pamukkale University

A survey form consisting of questions regarding the classification of practitioners and proposals for solutions to prevent or correct immoral applications was created by selecting five of the "ethical dilemma cases" developed by the two writers mentioned above. The survey was administered to 148 students who studied at the Department of Political Sciences and Public Administration and took the Political Psychology course. The survey form was designed in compliance with the Turkish public administration system, and the students were asked to answer "yes-no" questions and explain their answers. They were also told that the questions had no single true or false answer.

The conclusions made on the answers received from the students are as follows:

Responses for Case 1

"You are an elected mayor. A nice business center is planned for your district but the municipal budget is not enough to buy the planned land. A construction company which plans to build a shopping center near the land, states that it can buy the land and start the municipality's business center project, and in return wants to make the construction higher than the building limits set in the existing zoning plan. The offer is made directly to you. Would you present the offer to the district council without any delay?"

Those who responded "yes" to the first question were described as "the ethicist" or the persons who have "conduct/behavior based on ethics that promote the public good and public interest". Those who responded "no" to the first question were described as "the corrupter" or the persons who have "conduct/behavior based on no ethics". This evaluation was made in accordance with the following requirements: The city manager or the elected person should not decide alone in the face of ethical dilemma problems and hide information from the city council. In accordance with the principle of separation of powers, he/she should not take action by assessing the problem according to his/her personal moral values, but should present the matter to the city council (Steinberg and Austern, 1996: 17).

CASE 1	Number	Percentage (%)
Yes	51	(%) 34.46
No	97	(%) 65.54
No response	-	
TOTAL	148	

The fact that 65.54 % of the students responded "no" to the question supports the idea mentioned in the book that directly rejecting the offer of land "donation" would be a better option than introducing the issue to the city council. This would be due to the fact that the company states the "present" will be given in exchange for a zoning plan change, and the change will economically benefit the company but will reduce the property value of another person or company (Steinberg and Austern, 1996: 16,17). Indeed, those who responded "no" made similar explanations. The number of students who stated that "the application was not made according to the zoning plan so the offer must be rejected directly" is very high.

Considering the way of implementing the principle of separation of powers in Turkey and the lack of sharp divisions among the powers, the mayor introducing this "unfortunate" offer to the city council will most likely be subjected to severe criticism in Turkey. It is necessary to evaluate students' responses from this point of view.

Responses for Case 2

"You and Mr. Meriç have been close friends for many years. Both of you graduated from the same department of the same university, witnessed each other's weddings and your wives are good friends. Mr. Meriç and his wife have invited you and your spouse to dinner at your birthday for the past ten years, and this has become tradition for you. Mr. Meriç can afford these meetings because he owns one of the biggest infrastructure companies doing business in your province and he does over 10 million liras of business with the municipality in your province every year.

You have worked as an insurance agent throughout your business life. You were nominated and elected as a member of the municipal council three years ago. As of now, you have been appointed as the chairperson of the committee responsible for material procurement under the Municipal Assembly's infrastructure works. A few weeks later you have your birthday and Mr. Meriç called you and told you that he reserved a place for your birthday in a newly opened restaurant which is very popular in the city.

Would you accept the invitation?"

Those who responded "yes" to the second question were described as "the ethicist" or the persons who have "conduct/behavior based on ethics that promote the public good and public interest". Those who responded "no" to the second question were described as "the functionary" or the persons who have "value-neutral or relativistic conduct".

CASE 2	Number	Percentage (%)
Yes	110	(%) 74.32
No	37	(%) 25
No response	1	(%) 0.68
TOTAL	148	

In this question which concerns whether an elected or appointed official gives up his/her existing friendships and networks during their term in office, the behavior expected from the official is to avoid commenting and voting on a matter related to his/her friend when such matter is introduced to the city council. It will also be a good choice for the official to accept the dinner invitation, if necessary, share the bill with his/her (inviter) friend or pay for it to avoid any suspicions. Considering a possible/potential business partnership between the official and his/her friend, it is obvious that people who do not know their relationship or those who see their dinner photos/videos on the local media will get suspicious about them. However, as mentioned before, it is not necessary for the public officials to give up all their friendships and connections during their terms (Steinberg and Austern, 1996: 15).

Most of the students responded "yes" to the second question, stating that "friendship is one thing, business is another". They also emphasized that the dinner meetings have become a tradition, and stated that they would not mix up business life with private life. Twenty-five percent of the students responded "no", mostly putting forward a reason for their answers as, "I may be misunderstood, it (the dinner) is not an appropriate activity due to my status and duty". Some of the students did not approve of it (the dinner) in terms of morality or found it inconvenient in terms of tender process.

Responses for Case 3

"Every day for some time past, the police officers from two police patrol squads working in nearby patrol areas meet each other for a coffee break at a coffee shop located close to the intersection point of three police patrol areas. They usually consume tea, coffee and toast. You have been assigned to one of these police patrol squads. Your first day on the job, you offered to pay the bill but the coffee shop owner said "Sir, you have no bill to pay, it makes us happy that our government officials are in the vicinity."

Other officers leave without paying. Would you pay for the bill? "

The case is designed on whether the police officers pay the bill. Those who want to pay the bill are described as "the ethicist", and those who do not want to pay the bill are described as "the corrupter".

If the bill is not paid or it is regarded as a simple honoring, other public officials may also conduct such behaviors, which may result in widespread acceptance of such "honoring" and pave the way to legitimizing immoral practices. Civil servants receive money for the exchange of their services, not according to whether business managers are satisfied with them. It is also inconvenient to think that tea and toast do not cost much, so drinking tea and eating toast at a café without paying the bill is not a moral problem for the officers, because small bribery incentives may lead to more severe bribery cases in the future (Steinberg and Austern, 1996: 22).

CASE 3	Number	Percentage (%)
Yes	132	(%) 89.18
No	15	(%) 10.13
No response	1	(%) 0.69
TOTAL	148	

Almost 90 % of the students responded, "yes, I pay the bill", stating that no payment of the bill means accepting a "bribe". They also reported that the uniform did not grant privileges for them. It is remarkable that 10.13 % of the students responded "no", stating that "If this is their habitual order, then I do not interrupt it", "If I break their habitual practice that has become a tradition and make a payment, I will be ridiculous in the eyes of my friends. Therefore, I fall in line with them", "I do not make a payment only once" and "If this is the established order, then I should fall in line with the others".

Even if they are few in numbers, those who responded, "no, I do not pay the bill" have brought the community and its dominant opinion into the forefront in making their decisions. This reveals that thoughts of the masses, their acceptance and general functioning of the society can influence individuals' thoughts and behaviors extensionally and successively (Le Bon, 1997: 28).

Responses for Case 4

"Mrs. Emine is a valuable employee. This lady has been working with you in your municipality for years and you can rely on her and ask her to work hard and do overtime when needed. She is ready to help you with almost every problem, and knows how to deal with the problems distressing you. You owe her a great deal for this.

Recently, Mrs. Emine came to you and told you that for some time she has been "borrowing" money from the municipality charity fund, transferred it to her account and has manipulated the relevant data to hide her transactions. The amount of money she transferred to her account was not very large, 30–40 liras in general, and she always paid back the money after a while. However, she deeply regretted and needed to explain the situation because she was very uncomfortable. She should be fired according to municipality staff policies.

Would you fire her?"

In this case, Mrs. Emine committed two crimes, not one: illegally transferring money to her account and illegal manipulation of the fund data. The main crime is of manipulating the fund budget for her own benefit as transactions of "forging receipts and notice of delivery" may cause "similar" financial public losses in the future. Even though the crimes committed by Mrs. Emine seem to constitute a sufficient reason to fire her, it is not easy to find an officer who has her qualifications and characteristics. Therefore, the manager should only warn her instead of firing, and place her in another department in which she will not have any financial responsibility and authority (Steinberg and Austern, 1996: 21). Considering these explanations, those who wanted to fire Mrs. Emine were defined as "the functionary", whereas those who did not want to fire Mrs. Emine were defined as "the ethicist".

CASE 4	Number	Percentage (%)
Yes	78	(%) 52.70
No	69	(%) 46.62
No response	1	(%) 0.68
TOTAL	148	

The percentage of students who responded "yes" or "no" to the fourth question is close to one another. The response was "yes" by 52.70 % of the students, mostly asserting that "it is not important whether the amount of money transferred to her account is small" and "if the employee is forgiven for doing so, others may later behave in the same way". In addition, they also emphasized that the rules required to fire her. Those who responded "no" to firing Mrs. Emine mostly brought the concept of "qualms of conscience" into the forefront. Those who emphasized that Mrs. Emine regretted and then informed her manager about the situation, and those who thought that the problem would be solved by warning Mrs. Emine and following her closely after being assigned to another department, constitute the majority of the students who responded "no".

Responses for Case 5

"You had worked as an auditor for six years in the municipality, and you had controlled purchase contracts amounting to millions of liras. However, you left your job in the municipality and took a business proposal from a private company which was very difficult for you to reject, having worked for the municipality for six years. Due to developments beyond your control, your business in the new company has not gone well and you had to leave your new job and became unemployed for six months. Later, the same company asked you to come back as a consultant in order to

> recover the company. You accepted the offer, but you could work for only one week. Unexpectedly, a vacancy announcement was made for your old position at the municipality which you applied for and were recruited after passing through all necessary recruitment steps.
>   About a year later, your department made an announcement for a purchase of a commodity similar to those purchased previously. Your supervisor asked you to participate in the purchasing committee and you accepted. Shortly after that, one of the new owners of the company you had formerly consulted with, called you to find out how your business was doing. He asked about this procurement during the phone conversation and said that his company would participate in the tender. You told him you could not discuss this issue, and he did not persist.
>   Would you leave the purchasing committee?"

Although this case seems similar to the second case, there are differences between them. This case concerns a problem arising from a tender process and the relationship between the tender parties who previously worked together. On the one side, there is an official who works in the municipality making a tender, who is also a member of the tender committee, but worked as a consultant in one of the bidder companies and made a phone call with one of the bidder company managers about the tender process without making any detailed explanation or elaboration. However, the second case concerns the fact that "the behavior expected from the official is to avoid commenting and voting on a matter related to his/her friend when such a matter is introduced to the city council". In case 2, the official introduced the offer made by his/her friend to the city council, but in this case, even if friendship between the parties is not emphasized, the parties with a relationship dating back some time have come together in a procurement process. Beyond the friendship, the relationship between the chairman of the purchasing committee and the private sector representatives creates an inappropriate image in the procurement process. Therefore, the person in this case should resign as the chairman of the purchasing committee and withdraw from the committee altogether. But, in no uncertain terms, he/she should explain the reason of resignation and withdrawal as a "disagreement" without giving specific details so that it does not create any suspicion over the other committee members (Steinberg and Austern, 1996: 21, 22). In the light of this information, those who responded "yes" to the fifth question were described as "the ethicist", whereas those who responded "no" were described as "the corrupter".

CASE 5	Number	Percentage (%)
Yes	17	(%) 11.48
No	130	(%) 87.83
No response	1	(%) 0.69
TOTAL	148	

The percentage of students who responded "no" is interestingly quite high (87.83 %), and they mostly stated that there was no reason to resign in the case. In addition, a considerable number of the students who responded "no" stated: "Why should I resign? This is my job". There were some students from those who responded "no" stating that they would remain neutral and would not allow their old relations to affect their jobs. A few of the students who responded "yes" stated that it was necessary for them to resign and withdraw from their duties in the tender committee and did not want to be under suspicion.

6 Conclusion

In order for the state to play an active role in developing moral values and imposing them on society, and in order for the political structure to feel responsible for adhering to these existing rules, these norms have to be shaped in accordance with the principle of universality. A political mechanism should be set up to ensure this as an application. This study aimed to examine the moral tendencies of prospective administration officials in the face of the difficulties and problems they encounter and to draw conclusions from their moral tendencies.

We asked three case questions to students about "the corrupter" moral behavior. The results indicated that the abuse of public authority or title/status of being a public official in order to gain advantage and obtain a concession received a strong reaction by students stating its inconvenience. In addition, in the first of the two cases, which required a high judgment ability in a moral dilemma, students expressed their opinions about whether an illegal change on a zoning plan should be introduced to the city council, and a significantly high percentage of the students stated that it would be appropriate to discuss the matter in the council. Those who responded "no" were perceived as thinking that they took into consideration the way of implementing the principle of separation of powers in Turkish politics and that the offer should be rejected before introducing it to the council due to lack of compliance with the law. It is also important that the emphasis on obeying laws has a considerable effect on the way people behave. The second case concerns a purchasing tender process and the relationship between a public official and a company in which he/she had previously worked. It is remarkable that the percentage of students who responded "no" was significantly higher than the percentage of those who responded "no" in this case. A large number of those who responded "no" stated that the task they were doing was already included in their job definition, so there was no reason to resign and withdraw their duties in the purchasing committee. They also emphasized that they would not confuse business with other relationships. As stated by the students, although it would be appropriate for the official to resign and withdraw from his/her duties in the purchasing committee in order not to mix the previous relationship with his/her business, this study could not detect why students insistently emphasized on the counter-discourse and to what extent the guarantees that students give for not mixing business and private lives will reflect on the behaviors of these persons who are nominated to serve in public administration.

One of the cases questioning "the functionary" moral behavior is about whether to fire an employee who has spent years diligently, laboriously and altruistically doing their job but transferred money from the municipality charity fund to her account and forged official documents. In this dilemma, the percentages of students described as "the corrupter", "the functionary" and "the ethicist" were close to each other, but the "ethicist's" view that emphasizes on rules and the application of rules has drawn attention. It is also an important result that the number of those who responded "no" and thus gave an appropriate policy response to the case question which stated that these crimes did not require her to leave the job, is high.

However, these kinds of survey studies are often criticized because participants display idealistic attitudes and give the ideal response instead of their real thoughts. Therefore, it should be argued to what extent the quantitative data on moral behavior actually reflects the attitudes that are exhibited in real life. Thus, the present study results should be compared with the data from relevant applications in other fields, and accuracy of the responses given should be tested using cross examinations.

References

Ağaoğulları, M. A. (1993). Siyasal Ahlak ve Devlet. (pp. 283–289). *Siyasal Ahlak ve Siyasal Ahlaksızlık*. (der.) T. Alkan. Ankara: Bilgi Yayınevi.

Alkan, T. (1993). *Siyasal Ahlak ve Siyasal Ahlaksızlık*. Ankara: Bilgi Yayınevi.

Aytemiz Seymen, O.; Bolat, T. (2007). Kolberg'in Bilişsel Ahlaki Gelişim Modellinden Yararlanan Etiksel Karar Verme Modellerinin karşılaştırmalı Analizi. *Akdeniz Üniversitesi İİBF Dergisi* (13) 24–61.

Çileli, M. (1981). *Ahlaki Yargının Zihinsel Gelişim Psikolojisi*. Ankara: Ankara Üniversitesi. Yayınlanmamış Doktora Tezi.

Ferrell, O. C.; Gresham L. G. (1985). A Contingency Framework for Understanding Ethical Decision Making in Marketing, *Journal of Marketing* 49(3), 87-96.

Güngör, E. (1993). *Değerler Psikolojisi Üzerinde Araştırmalar*. İstanbul: Ötüken.

Jones, T. M. (1991). Ethical Decision Making by Individuals in Organizations: An Issue Contingent Model, *Academy of Management Review* 16(2), 366-395.

Mercin, L. (2005). *Piaget ve Kohlberg'in Ahlâki (moral) Gelişim Kuramlarının Özellikleri ve Karşılaştırılması*, SBArd, 5 73-86. http://www.akader.info/sbard/sayilar/2005Mart/73-86.pdf erişim tarihi: 11.10.2010.

Onur, B. (2000). *Gelişim Psikolojisi: Yetişkinlik, Yaşlılık ve Ölüm*. Ankara: İmge Yayınları.

Rest, J.R. (1984) *The Major Components of Morality*. New York: Wiley

Steinberg S.; ve Austern D. (1996). *Hükümet, Ahlak ve Yöneticiler*. Ankara: Türkiye ve Orta Doğu Amme Enstitüsü Yayını.

Trevino, L.K. (1986). Ethical Decision Making in Organizations: A Person-Situation Interactionist Model, *Academy of Management Review*, 11(3) 601-617.

Bilge Ünal

Cultural Dimensions in Migrant Literature Stereotyped Usage in Authentic Stories

1 Introduction

Systems of various societies are being observed, examined, compared, confronted, and criticized. The intention of such scientific piece of work is based on a qualitatively and quantitatively high life expectancy of every single individual and external factors, pulling an intact culture out of its system which must be inserted, so that the system of this society takes its course perfectly without any compromises. That's why Groth's ascertainment "Each living thing is linked to the objective world" (Groth, 1960: 126 ff) is a good beginning to consider cultures more closely. The individual must develop its possibilities in communication with reality, nature, society, and culture. Langenbucher (1964) assumed that there is an aspiration in the inner self of each human being, which creates a desire or rather a dream world out of an associative partiality and a divined universality of rational, emotional, and mental self-development.

1.1 About German History

After the Second World War in the 1950s, the German economy expanded and a lack in human labor was seen in the industry. The high demand on the job market should be covered for a not quite unratable but limited time with the help of foreign manpower also known as 'immigrant workers'. The primarily purpose degenerated itself to migration of these *exported labor*. Later, their families and relatives came after them. It was expected from the society wanting immigrants only as labor, to fit in the culture but not to settle down. Over the years, that adjustment on culture has led to a modification of values and norms through to an alienation of the migrants' culture and country of origin. The migration, the unfamiliar culture, low social and political state, language and conversing barrier have strengthened uncertainty and "being lost" of these migrants. The immigrant experienced his self-expression in a small part and classified his environment and outside world with his attitude towards life, unconsciously as a "counterworld" (Term from: Groth, 1960: 126)". The literary group Südwind, which was established in 1980, wanted to attract attention and had also a protest language that constructed the "Südwindgastarbeiterdeutsch" (Hlavinova, 2013: 15). With the third generation of the authors like Selim Ozdogan, linguistic variabilities such as practical sentence patterns, limited vocabulary, not frequently used prepositions or conjunctions have been added (http.- 7). To Hofmann, language is a tool which can diminish a conflict caused by an encounter of various cultural dimensions: […] "With this in mind, intercultural responsibility is not to mix up with the ability of understanding everyone and everything. No, the point is to accept strangeness - Literature is able to contribute to that as well (Hofmann, 2006. 8; Hlavinova, 2013: 15)". In the 1950s, Germany has developed itself from the Marshall plan (1947), the monetary reform (1948), and from the foundation of the Federal Republic of Germany (1949) upward to a mentally pragmatic and efficient economic miracle. A great significance was attributed to the German Mark labeled as 'Made in Western Germany' and to the increase of the gross national product. An involvement of the western zones as FRG in NATO was made possible by the political and social consolidation, whereby the Cold War (Superpower USA and the Western powers NATO against superpower Soviet Union, East block/Warsaw Pact, first high point: break out of the Korean war in 1950) also had an engraving impact on all European countries (Ivers/Witt, 1978: 127–188).

The consumption and success displaced the upcoming work-up of the NS-past and the Holocaust. Philosophers and sociologists confronted themselves with the past. The 1950s were the heydays of the radio play, in which Günter Eich's dreams were first sent on the radio in 1951 (http. -7) (Eich, 1951). Paradigmatic movies and German music, which should be reflecting the longing for an ideal world, were produced at this time. There wasn't a homogenous conception of mankind, yet, the recognition has been bundled up with the first article of the Basic Law in the German Civil Code saying that 'Human dignity is inviolable'. Towards the end of the 1950s, authors adopted a critical attitude towards the conservative policy. In the use of language, there was a tendency to an 'unadorned-meager' speech (Stark, 2013: 33).

With the contract of recruitment "The contract of recruitment was signed with the following countries: Italy (1955), Spain (1960), Turkey (1968), Morocco (1963), Portugal (1964), Tunisia (1965) and Yugoslavia (1968) (Hlavinova, 2013: 9)" the labor migration was getting it on. In the year of 1956, the first guest-workers, in October 1961, the first Turkish guest-workers (Korte, 1983: 21; Kocadoru, 2003: 5) came to the Federal Republic.

1.2 About Generations between Migrant Writers

Turkish authors, who also immigrated to Germany for various reasons such as Yüksel Pazarkaya, Aras Oren, Güney Dal, and Bekir Yıldız representing the first generation of the migration literature, corporate short stories, novels, fables, and poetries about the experiences of the guest-workers in their stories and reflect those as historical evidences, more specifically as documents (Kocadoru, 2003: 4). The "Structure of the societies, the cultures and the worlds" such as Langenbucher (1964) on Maslow's Hierarchy of Needs defines it interpretively:

> [...] Only in case of fulfillment higher desires are more likely to be visible than those on a sacle of the lower ones. The longing after safety begins so to emerge, when hunger and thirst are satisfied; the wish for love should warranty the safety, and the reputation should then appear as a problem, when the longing for love is accomplished too. Once the desire for reputation is satisfied, the desire for self-actualization should adjust itself (translated from: Langenbucher, 1664 a. a..O.: 171; Maslow).

1.2.1 First Generation of Migrant Authors

Unlike the conditions of the first generation of migrants, especially basic needs are conveyed in migrant literature. As requirements do not nearly cover the basic needs of migrants, they are also seen as a reason for revealing Germany in a negative way in the first migrant literature. The migrant literature, literature between two cultures, literature about seasonal workers in immigration countries, foreign literature, literature of German-speaking Turks, Turkish-German literature, literature of minorities in Germany (covering the 1970s and 1980s), multicultural literature, literature of a major culture i. e. (translated and comparable with Kocadoru, 2003: 6 – articles and lectures Istanbul University, 2003) following after Kuruyazici (Kuruyazici, 2001: 19; Kocadoru, 2003: 10), immense solitude, homesickness, alienation, adjustment problems, a position as outsiders, being socially pushed in the country of origin, and financial shortage of migrants in a world for them being seen as a strange one. This generation is confronted with "many and not sustainable humane problems". The "Literature of consternation" - "acı çekenlerin yazını" (Özoğuz 2001: 202; Kocadoru, 2003: 10) is captured by Oraliş (2001: 40) clearly through the attempted explanation of the experienced, not at all to experience. Individual testimonies (Protagonist) of tragic events, longing, homesickness, renaming, and a whole new meaning of one past and in comparison with the existing external conditions of immigrants, to be made understandable for the readers in order for them to be able to empathize and to put themselves in the shoes of seasonal workers. These are authentic works with stereotyped properties which are seen as crucially important and carried out in a detailed way. Wenzel (1978) gives a clear transcription for the term 'Stereotype':

> A stereotype is the verbal expression of one of the social groups or an individual person as their members of a directed movement. It has the logical form of a general statement

which, in an unjustly manner, with emotional-judging and a normative tendency to attribute or agree to a class of people with certain characteristics or behavior (compare: Wenzel, 1978: 28).

1.2.2 Second Generation of Migrant Authors

The second generation of migrant literature begins in the mid-1980s following the wave of Europe. They are victims from expectations of a new cultural dimension, a purposed union of neighboring states, a cultural fusion, an attempt to enable an integration of guest-workers back to the country of immigration, a project cancelled for financial reasons (Oraliş, 2001: 58; Kocadoru, 2003: 24). Especially the life of a guest-worker, the difficulty of a cohabitation, and the cultural strangeness. The 'foreign authors' were, in most of the cases, the ones who immigrated, who were intellectual, studying, or political emigrants. One doesn't speak of guest-workers anymore. Even the second generation of foreigners starts writing, too. This group still familiarize themselves a bit with the subject of guest-workers, but now the identity problems are covered in the foreground (http.-6)". The second generation are represented by Zafer Şenocak, Levent Aktoprak, Zehra Çırak, Renan Demirkan, and Nevfel Cumart. They write about how migrants between two cultures feel constricted, a cultural separation with a statelessness searching for a third hideaway denounced with helplessness, tumbling, floundering down a slope into an endless loss of existence trying to find an identity.

1.2.3 Third Generation of Migrant Authors

The current so-called third generation of migrant literature has strongly changed itself. The new era no longer includes problems of a suppressed immigrant folk. The 'Literature of Guest-workers' is way ahead, claims Schuhmann (http.-8); he even says that the new generation with background of migration has finally reached normality. Authors tackled the issue of migration scientifically. Migration has become a sub-area of philosophy (Kocadoru, 2003, S. 30). Selim Özdoğan, Orkun Ertener, Kadir Kurt, Şener Saltürk, İsmet Elçi, Feridun Zaimoğlu, and Emine Sevgi Özdamar belong as authors belong to the young literature (compare with Kocadoru 2003: 32/33).

1.3 About the Author Bekir Yıldız

Bekir Yıldız (1933; † 1998), born on 3rd March 1933 in Şanlıurfa, was a Turkish- and Kurdish-speaking author, who wrote his stories in the Turkish language. He started his career after finishing the training as a printer and going to Germany as a worker (comp. Yıldız 1971). He completed his Military Service in the year of 1957 in Eskişehir and there, he gets to know his wife Güler and gets married to her that same year (Baskak 2008: 1). After working as a typesetting teacher for a short time, he migrates to Heidelberg in 1962, a city which is presumed to be the center of a printing industry in Europe (1962–1966) and works as a Turkish guest-worker in a printer-manufacturing factory in which he gets his family (Ms. Güler, his daughter Vildan, and his son Yüce) to join him shortly afterwards. Then, he publishes his book including his own experiences in Germany called "Turks in Germany" (gr. Die Türken in Deutschland) (http-1) in his own printer which he established in Istanbul. The positive image of Germany in the Turkish Literature was first (http.-2) destroyed by this work (http.-11). In 1974, the book *Alman Ekmeği*, (comp. "German Bread"; gr. Deutsches Brot) was written in the Turkish language and also published in the magazine called 'Kürbiskern' (http.-2). The author belongs to the most readable, sociologically oriented writers. His works 'Kara Çarşaflı Gelin' (ger: schwarz umhüllte Braut) and 'Bedrana' were made into a film.

In the essence of his writing, a reality check takes place whereas he always has the final say on real-life ventures. The reader must always keep himself on a realistic level and question reality. Those stories being told by people who traveled to Germany in order to find work have a significant position in Yıldız' works. In these short stories not only problems of the working class have retained like in Sahipsizler (Yıldız 1971), he also mentions how the system in Germany is leaving human rights out of consideration and it is also leading to adaptive difficulties in familiar customs and cultural habits in a strange place as well as mental and psychological wounds as well as to an alienation of those people. "Sahipsizler", "Celb", and "Maria Otuz İki Yaşında" are only a few examples of these short stories. "Motorize Köleler" through its narrative style and visual narration has a special place in Bekir Yıldız' stories. Uturgauri (1989) is seeing Yıldız' perceptions of experienced reality as a narrative style which basically degenerates itself into a tragedy in order to increase his nation's life quality as he remembers them not to lose sight of reality, barriers of the Traditional, he exclusively aims to enlighten immigrants.

> The authenticity happening in the life of Bekir Yıldız is going to bear tragic situations at any time, tragic situations affecting into daily life are getting usual. (...) Analyzing the reasons of tragedy is not an aim for a passionate author like him; the development of public's awareness, morals which prevent the rise of comfort standards, remains of feudalism-tribes in a society being in existence, are all tools for human relations to reveal. (Uturgauri, 1989: 190)".

2 Hofstede and His Thesis in Migrant Literature

Bekir Yıldız expresses his own attitude by praising the positive development of Turkish stories, and he sees the very creativity in a way that content, expressive form, repetition of reality, societal and traditional backgrounds picked up individually and the future prospect is merged to the story:

> In order for Turkish literature to show an honorable development whether it's through abstract formality or vulgar naturalism if needed, it is going to happen. The problem is to catch the Turkish being under Turkey's conditions and to explain the importance of its style. The real creativeness goes as follows: To combine the native and competent essence. The factual plotline before mentioned, is not a midway of formalism and naturalism. A an attitude, it's way different and apart than from the other two. (...) it's open for future. (translated from: Yıldız, a. g. e.: 98).

2.1 Individualism versus Collectivism

Collectivity instead of Individualism is clearly fetched through the large class distinction between guest-workers and natives. According to Hofstede (http.- 9, 2014), Individualism refers to the sovereignty and affiliation of people in a society. In "Sahipsizler" (Yıldız, 1971: 4), Spanish are supporting Turkish guest-workers by accompanying the corpse in a coffin to the graveyard. Translated:

> How did he die, Amigo? [...] - "I don't know!" he answered slowly in a carefully silent voice. "Let's go with them Amigos... On a day like this there can't be a separation between Spanish and Turks...' Those Spanish people, they were impetuous. Plus, they were warmhearted and amiable. [...] (translated from: Yıldız, 1971: 4).

Hofstede (2001, http. - 5) explains as in a culture of "high Individualism" a focus on individual success and personal rights are required and these basic conditions of guest workers are missing for an Individualism, it leads to a collectivity, a particular unity and group membership among migrants and their families whose needs and aims are more important than those of individuals, the migrants so to say. Other features include the poor economic relations where these migrants come from, and they already bring a collectivity with them which is a conditional culture of origin as they had accepted this migration for their families and their well-being in the first place.

In "Demir Bebek" by Yıldız (1995) for instance, the perception of feelings, the helplessness of the migrant worker families were made perceptible with the help of a simultaneous technology whereby the most various pictures replace themselves in a rapid succession, and those time periods which are crowded together are suggested. Those techniques especially preferred by urban authors such as Alfred Lichtenstein (1962), and Jacob van Hoddis (1958) "without unnecessary reflection" (Vietta & Kemper 1975: 33), reflect sociological and psychological requirements, the perceptions and reality, the chance of outer and inner impressions of people living in foreign metropolitan habitats (Simmel, 1975: 34).

2.2 Uncertainty Avoidance Index (High versus Low)

Doubtful situations and changes can, according to Hofstede (2014), be seen as a threat. By Hofstede a high prevention of uncertainty in a culture means for human being major anxiety states and more situations of stress on obscurities, changes or situations without any rules (http. -5). Rules of conduct are supposed to reduce such anxiety states or more specifically to prevent them. In Hofstede's view (2914), a slight avoidance of uncertainty means that in a culture it is possible to solve obscurities and changes easily in case the culture is practical, is good with changes, and utilizes rules as little as possible.

In his novel *Muhteşem Gündoğdu*, Aras Ören brings out a shocking change of the loss of books out of a book store which was emptied and closed from one day to the next which takes away the reader's hope to solve problems (Ören, 2011: 89). He is trying to suppress his helplessness and uncertainty with the knowledge he gained from books he was reading and the souls and adventures he was putting himself into, only to escape out of the misery into an intact world. He is disappointed because with all of these books which stood for souls and lives, he wanted to construct a Tower of Babel (Groth 1960: 126) itself, from a tower of salvation, a salvation of one's own personality crisis into a self- and understanding of reality with the aim of a later healed dream world.

2.3 Masculinity versus Femininity

According to Hofstede (2914), Masculinity (MAS value) is related to whether in a society or culture values such as material success, aggressiveness, ambition and competition, and classic division of roles are distinctive as stereotypical masculine. The word Masculinity is, according to Yıldız in his work called "Alman Ekmeği" (German Bread), indicated as the following: "My son's yellow, fine moustache is sweating, shivering. For all of my sons' moustache are sweating. (Yıldız, 1974: 4)" Are only men and boys working? Where are all the women and girls?

In Bedrana, tradition is equated in a men's society with injustice, treachery, and egoism. Translated: "Naif wanted to complete his plan he staged for so long on this very night before tomorrow comes. With a kick, he knocked over the pillows which were under his wife's feet and wanted to fulfill his plan this night (Yıldız, 1971: 20)". Even though he knows his wife is innocent, he wants her to die without having to go to jail himself. He is afraid of a loss of face but his actions are devious murder which are justified through trouble and death. He is saving himself and his honor towards those of the same sex since the rape victim as woman and as an evidence for the weakness of men must be kept out of the way and stamped as guilty, so that the one using violence coming out of a male-dominated society can be traditionally protected.

In a feminine culture, more emphasis is put on regard, on empathy, and on altruism than on status and competition and aims non-hierarchical communication on an equal level when it comes to a low-power distance. Since *creativity* carries the risk of uncertainty (http.-7) with itself, it's not permitted in cultures having high-power distance and clear statements.

2.4 Power Distance Index (High versus Low)

The shocking studies of Yıldız do not give the chance of exploring other opinions to escape from reality for in reality basically always tragedy takes its course and compared to the grotesque, it is seen as a pragmatical development. The existing disastrous situation, the warm-heartedness of these desperate people, a verbal and wide range of regional accents, natural events, sociological circumstances are showing the gradual change of these people.

Based on self-criticism filled with feelings of guilt, Yıldız drops a hint on his own cultural dimension. Instead of confronting himself with the problems of his homeland and finding a solution, he makes a decision for the simpler way, by putting himself like a parasite on Germans, fleeing from the oppression which due to high-power distances "in one's own culture" became a feature of life itself. Here, a proof for Bekir's evidence from his work 'Alman Ekmeği - Ekmekle Körebe oynayanlar' (germ.: Deutsches Brot - Die, die mit Brot 'Blinde Kuh' spielen):

> The growling stomach starving in mother state, yet here saturated stomach, that's how it started. There, on these tracks where the train had stopped. We've already got out, 32 relatives. From thousand, two thousand, maybe three thousand kilometers of distance, even the invisible but always those pain of lashes being felt on my back has dragged me here. That was the destiny of my shift. I was a slave. But what kind of a slave? The one being snatched from his fields and taken up as freights on ships with weapons and lash prints and kilos of sold a slave? O a tortured kind of slice, who escaped from the torture and humiliation of his master? Or a slave, who, instead of being able to solve most of his problems in his own country, took the easier way out and assessed? Not decisive of what kind of slave I'd like to be, I do come from a tribe, from a nation, in which 'Blindman's Buff' was played with bread (translated from: Yıldız, 1981: 2).

How strong social inequalities between superior and inferior employees are in place is showed according to Hofstede's study (2914) of the degree of power distance in this culture. Although Germany shows a lower power distance in this study, high distances between guest-workers and natives are listed in Yıldız' short stories. Translated:

The local factory manager just said a short while ago, that there is no need for a discussion. I will announce the result briefly: The dead person is going to be burned to death, as the place on the graveyard in this town is full. It's out of the question to build a new graveyard. If you want to know the reason for that, around those big cities there will still be factories to be build. We are already burning our corpses and that's for a long time now. What else can there be more natural.... (translated from: Yıldız, 1971: 8).

The tragedy will be included in everyday life to a habitual procedure. At the same time, with this extreme reality it shows the unequal distribution of power in a familiar culture dimension in the form of terms such as 'Töre'.

Exactly these points are undertaken by Emine Sevgi Özdamar (2013) in her novel *Life Is a Caravansary – It Has Two Doors – I Came Out from One – I Stepped Out of the Other One*. She describes the helplessness facing poverty, that feeling of being restricted of norms and rules in one's own society, masculinity, unreal power distribution, uncertainty and frightening changes of lower compliance values limited on the part of one's family, religion and traditions. The expectations of the society, the family, relatives, environment, the high self-control, leading in psychological extends to serious diseases cause the first person narrator to making a decision. In a promising way, she is choosing the getaway with the "Black Train", which symbolizes a travel into the unknown, but does not impose any hints and for the first time represents a travel into an emancipation, away from impeding conditions in one's homeland, into the freedom 'of luck'. "I got on the train to Germany, as many women also did, there was only one single man who got on the train, who was the traffic controller [...] It was a train full of prostitutes (Özdamar, 2013: 379)", to Germany.

2.5 Indulgence versus Restraint

As to Hofstede (http. -5), a high compliance value or a high Indulgence allows fulfilling basic needs and wishes of the individual to satisfy unhindered or to fulfill independently when compared to Resistant-value (IVR). The importance of luck and control over one's own life are detected and supported. However, in societies with high extent of control (lower IVR-Value) needs are being oppressed and social norms delimited the freedom of the individual. Moral discipline has, in these cases, a great level of importance, and there is a tendency to pessimism. This tendency is seen in Aras Ören's (2011) novel *Muhteşem Gündoğdu*, in which the narrator is morbidly attached on his past and habits and in these empty expectations from the past, homeland and traditions he enhances into it with emotional outburst until he frustratedly meets Kafka in his daydream (Ören 2011: 94).

2.6 Pragmatic versus Normative

According to Hofstede (2014) societies with a long-term alignment encourage to entail larger investments and be more economical. Furthermore, endurance and attention are leading to rewards. The members of a society have solid, clear, and social positions. Older people must be respected and relations must be appreciated. These cultures with high long-term orientation values (LTO = Long-Term Orientation) tend to traditions on modern contexts to be adjusted. Family and relations occur before economical values. Societies, which are geared last minute, do respect traditions, but encouraging expenses and immediate profits. In this society, status does not matter and relations are only seen as essential when one can benefit from them, as profits of this society are on the top of the pyramid's aim. In the short stories of Yıldız, migrants' rights to live are taken away. In a foreign land, those people are labor workers without any rights in an economical society fixated on profits. In "Ekmekle Körebe oynayanlar" (engl. "Those playing blindman's buff with bread") Yıldız (1974) indicates on this profit-fixed society. Translated:

> Very young, with her small hands which were at a playing-, reading, learning age, there she opens the garage door. They wanted to give support. They've earned minimal money since their work was cheaper than plastic toys. Suddenly, the streets of Walldorf were surrounded and crowded by children of their homeland as if the petrol storage was filled with the blood of each child." (translated from: Yıldız, 1974: 3).

With the verbalization of child's blood in fuel especially profit leads here to a criticizing reference on an unrealistic distribution of power which leads to a non-compliance of the human rights in the German Civil Code. Here, it is essential that we add the fact that writers of the first generation draw a positive image of homeland and unlike that, a negative image of the immigration country (Tekinay, 1983; 207; Kocadoru, 2003: 12). It is questioned whether the condition from the perspective of a viewing angle are real or not, because who leaves an intact life to be exposed to requirements in a foreign country, in an unknown culture, far away from trust?

3 Conclusion

In those stories of Bekir Yıldız, psychological excitement of a migrant in absolute reality appears as mechanically causal laws which are subjected functions. The author is seeing each one of them as an object of society-related forces which, here, take the freedom of Individuals in this case that from the migrants, and revamp those. The reality seems different and becomes non-human and cruel. Destinies of those migrants are being displayed in a provoking grossness. The usage of different stylistic levels (sophisticated, dialects, group language) and the function of the language as an instrument of oppression, as means of expression of social differences between immigrants and natives, as a mirror of inhumanity, as a horizon of experience and the psychological condition of migrants is masterfully used. The connection of the series of scenes with simultaneous technology, Characterization of the people through immediate action, non-action and behavior in critical situations and environmental factors show the form of the issue effectively to advantage. Consequently migration to Germany has declined overall, as the third generation of immigrants are living, working and writing in Germany, and as the mother tongue is more and more being the German one, their literature being produced in the 1960s has finally arrived at some point he calls "Normality" (http.-8) and it's also according to Schuhmann's (2006) integral component of a German-speaking literature without there being a sign of the eccentrically and marginal.

Moreover, he determines that, for several years, in literature, "authors of several cultures" whose mother tongue was not German, rarer and rarer (http.-8). The sinking tendency shows us that the era of the Turkish migrant literature comes to an end, which indicates that a culture melting has taken place, and already existing problems are in change and that from a few aspects such as anti-foreigner attitudes, unemployment, and so on are being repressed.

References

Radio Programs

Eich, G. Träume. Regie: Fritz Schröder-Jahn, Musik: Siegfried Franz-, gesendet vom Nordwestdeutschen Rundfunk Hamburg am 19. April 1951.

Journals/Periodicals

Baskak, S. (2008). Bekir Yıldız'ın Öykücülüğü. Çukurova Üniversitesi Sosyal Bilimler Enstitüsü Türk Dili Ve Edebiyatı Anabilim Dalı aus: Yıldız, Bekir (Mayıs 1970), "Kendileri (Otobiyografileri)," *Papirüs Dergisi*, İstanbul: Gün Basımevi, S. 1, 46–47, S. 98.

Korte, H. (1983). Die Bundesrepublik zwischen Einwanderungsland und Saisonarbeiterzug. Zum Thema - Wird Mehmet zu Michael?, Die Integration ausländischer Mitbürger und Hilfen zur Verwirklichung, Heft 5, Düsseldorf, S. 21.

Kuruyazıcı, N. (2001). Almanya'da oluşan yeni bir yazının tartışılması - Gurbeti Vatan Edenler- Almanca yazan almanyalı Türkler, Karakuş, Mahmut/Kuruyazıcı, Nilüfer, (Kulturministerium/T.C. Kültürbakanlığı) S. 19.

Oraliş, M. (2001). Gurbeti Vatan Edenler. Almanca yazan almanyalı Türkler, Karakuş, Mahmut/Kuruyazıcı, Nilüfer, (Kulturministerium-T.C. Kültürbakanlığı), S. 40.

Simmel, G. (1975). Die Großstadt und das Geistesleben. In: Die Großstadt. Jahrbuch der Gehe-Stiftung Dresden 1903. Hier zit nach: S. Vietta/ H. G. Kemper: Expressionismus, München (Deutscher Taschenbuchverlag).

Uturgauri, S. (1989). Türk Edebiyatı Üzerine, Cem Yayınları, Istanbul, S. 190.

Thesis

Hlavínová, Jana (2013): Diplomarbeit überbold> /bold>Selim Özdogan's, "Die Tochter des Schmieds" und "Heimstraße 52". Masaryk Universität Brünn, Philosophisch Fakultät, Institut für Germanistik, Nederlandistik und Nordistik, S. 9, 15.

Books

BGB, (1900). Bürgerliches Gesetzbuch (seit 1.1.1900 in Kraft) Enthält die heute gültigen Bestimmungen über Rechte und Pflichten des Menschen im Privatleben (Zivilrecht).

Groth, O. (1960). Die unerkannte Kulturmacht. Grundlegung der Zeitungswissenschaft, Berlin, Verlag Walter de Gryter, Bd. 1, S. 126 ff.

Hoddis, J. von (1958). Weltende. In: Weltende. Gesammelte Dichtungen. Zürich: Verlags AG "Die Arche", Peter Schifferli.

Hofstede, G. (2001). Culture's Consequenses. Comparing Values, Behaviors, Institutions and Organizations Across Nations. Thousand Oaks: Sage.

Hofstede, G. (2017, 6. Auflage). Lokales Denken, globales Handeln. Interkulturelle Zusammenarbeit und globales Management. München: Beck.

Ivers, U./ Witt, G. (1978). Geschichte-Politik-Technik-Wirtschaft-Sozialkunde In: Der Hauslehrer. 7. Auflage, Hrsg. Theodor Müller - Alfeld, Südwest Verlag GmbH & Co KG München, ISBN 3 517 00529 0, S. 127–188.

Kocadoru, Y. (2003). Geçmişten günümüze Almanya'da almanca yazan Türkler ve Emine Sevgi Özdamar, Rema Matbaacılılk A.Ş., Istanbul, ISBN 975-288-745-7.

Kuruyazıcı, N. (2001). Almanya'da oluşan yeni bir yazının tartışılması. Gurbeti Vatan Edenler. Almanca yazan almanyalı Türkler, Karakuş, M./Kuruyazıcı, N. (Kulturministerium/T.C. Kültürbakanlığı), S. 19

Langenbucher, W. (1964). Der aktuelle Unterhaltungsroman, Bonn: Bouvier.

Lichtenstein, A. (1962). Die Dämmerung In: Gesammelte Werke, Zürich: Verlags AG "Die Arche", Peter Schifferli.

Ören, A. (2011). Muhteşem Gündoğdu. Istanbul: Alfa-Everest Yarınları, S. 89, 94, ISBN 978-975-289-929-2.

Özdamar, E. S. (2013). Das Leben ist eine Karavanserei - hat zwei Türen - aus einer Tür kam ich rein - aus der Anderen ging ich raus. 1992 erschienen, 8. Auflage, Köln: Verlag Kiepenheuer & Witsch, S. 10, 379 ISBN 978-3-462-02319-0.

Stark, (2013). Deutsch - auf einem Blick! Epochen der deutschen Literatur. Stark Verlagsgesellschaft mbH & CO. KG, Freising, S. 32.

Tekinay, A. (1983). Langer Urlaub. - In zwei Sprachen leben. - Berichte, Erzählungen, Gedichte von Ausländern, Hrsg. Irmgard Ackermann, DTV, München, S. 207.

Wenzel, A. (1978). Stereotype in gesprochener Sprache. Reihe 1: Linguistische Sprachen, Max Hueber Verlag, München, Band 13, S. 28.

Yıldız, B. (1971). Sahipsizler. Everest Alfa Yay. 2012, İstanbul, S. 4, 20, ISBN 978-605-141-181-1.

Yıldız, B. (1974). Alman Ekmeği. Everest Alfa Yay. 2011, İstanbul, S. 1-4 ISBN 978-975-289-858-5.

Yıldız, B. (1995). Demir Bebek. Istanbul, Cem Yayınevi, 10. Basım, S. 5-13 ISBN 975-406-532-2.

Internet Sources

https://de.wikipedia.org/wiki/Bekir_Yıldız, (6. March 2018).

http://deacademic.com/dic.nsf/dewiki/152031, (6. March 2018).

http://focusmigration.hwwi.de/typo3_upload/groups/3/focus_Migration_ Publikationen/Laenderprofile/LP01_Deutschland_v2.pdf S. 2., vgl. OECD http://www.oecd.org/berlin/ 35796774.pdf S. 15., vgl. Bundeszentrale für politische Bildung.- http://www.bpb.de/politik/ hintergrund-aktuell/68921/ erstes-gastarbeiterabkommen-20-12-2010 [9. 7. 2013]., SCHEU: Migrace a kulturní konflikty, S.281. (26. February 2018)

http://webdoc.sub.gwdg.de/ebook/dissts/Siegen/Photong1996.pdf S. 36. Niedersächsische Staats-und Universitätsbibliothek Göttingen (26. March 2018).

Hofstede, G. (2014). http://www.typentest.de/blog/2014/08/hofstedes-kulturdimensionen-wie-uns-die-lan-deskul-tur-beeinflusst/ (18. March 2018).

Deutsch-Türkische Medienbeziehungen 1999–2009, https://books.google.com.tr/ books?isbn=382604522X. (29. March 2018).

Eich, Günter (1951). Träume. ARD-Hörspieldatenbank: www.hörspiele.dra.de

Schumann, Thomas (2006). https://de.qantara.de/inhalt/neue-tendenzen-in-der-migrantenliteratur-der-gast-der-keiner-mehr-ist (20. March 2018).

Hande Ünsal
Admission of Foreign Real Persons in Turkey

1 Introduction

The phenomenon of travel which is almost as old as the human history has increased considerably in the last centuries parallel to developments in transportation technologies. This increase carried the travel[1] and the related issues, among the important topics of both domestic laws and international law.

Individuals' travelling in the country which they are the nationals of are often considered within the scope of human rights (Aybay, 1979: p. 80; Gürsel, 2017: p. 463). However, international travel is not readily evaluated under the scope of human rights (Seviğ, 1977: p. 368). According to an agreed principle of international law, every state has to admit its nationals entry into its territories; whereas, it does not have an obligation regarding the foreigners (Çelikel, 2018: p. 81). Under the principle of territoriality, states freely determine the conditions of foreigners' entry into their territory (Özkan, 2007: p. 410). In this context, the state has the authority to obstruct the entry of foreigners into its territory, to constrain it or to bind it to certain conditions. States' sovereignty is limited only by international agreements, and apart from that, there is a wide range of difference between the approaches adopted by states in accepting foreigners to their territories (Aybay, 1979: p. 81). However, it is expected from the states to use their discretion in a non-arbitrary fashion and regulate the procedures and conditions for the entry of foreigners into the country objectively by law (Doğan, 2017: p. 37). The points that foreigners can enter the country, the documents required to present during entry, procedures, the visa requirements for entry and the legal ways that foreigners may apply due to the refusal of entry into the country should be subjected to objective regulations in order to provide legal security and prevent arbitrary treatment. Nevertheless, fulfilling any requirement that is determined by the regulations on entry into the country does not grant absolute right of entry into the country for the foreigners (Çelikel, 2018: p. 81; Doğan, 2016: p. 38); the competent authorities reserve the authority to refuse the entry of a foreigner

---

1 Travel can be made on the basis of different reasons, it can be result of an independent choice or necessity. Unlike the term "immigration", the term "traveling" is often used to mean "a relocation without the purpose of settlement". Nevertheless, each migration movement also begins with a travel as well.

into the country (Doğan, 2016: p. 38). Thus, entry of foreigners into the country emerges as one of the areas where the states have the greatest discretion.

Turkish Law regulates access to the country as a constitutional right for nationals. Under Article 23 of the Constitution of Turkish Republic[2] (The Constitution), a national cannot be deprived of entry into the country. As a result of this, it is not possible to ban or refuse the entry of a Turkish national in Turkey. On the other hand, like many other countries, the Turkish Constitution limits the scope of absolute access to Turkey to its nationals; the foreigners will be able to enter into Turkey in the framework defined by the law.

Our study focuses on the foreigners' entrance conditions to Turkey. In this context, the issues related with foreigner's entry in Turkey such as the documents to be presented in border check points, visa requirements and the check in the border gates will be evaluated. In our study, the term "foreigner"[3] is used to denote other state nationals and stateless persons in accordance with the general concept of Turkish law. However, regulations concerning the entry conditions of foreigners who are under international protection, refugees and/or conditional refugees are excluded from the scope of the study and will be dealt to the extent that is required by the course of the study.

2 Foreigner's Entry into Turkey

For a long period conditions and rules concerning the entry of a foreigner to Turkey is regulated by the Passport Act Law no. 5682[4] (Passport Act), adopted in 1950. The Law on Foreigners and International Protection[5] (LFIP) Law No. 645, adopted in 2013, has abrogated some provisions of the Passport Law and introduced new regulations. Nevertheless, some basic principles and provisions contained in the Passport Law are still in force (Çelikel, 2018: p. 81). Therefore, with regard to foreigners entering in Turkey, the Passport Law is and LFIP applies in conjunction.

---

2   Law no. 2709 dated 7.11.1982, Official Gazette: 09.11.1982-17863 Duplicate.
3   In Turkish Law, the term "foreign" (*yabancı*) is used to express a connotation involving foreign nationals and stateless persons. Even though it is more appropriate to engage the word "alien" to meet this meaning in English language, the term "the admission of foreigners" appears to be used more frequently in the literature than "admission of aliens". This study will keep up with the general tendency in the literature and will also include the term "foreigner" to refer to other state citizens and stateless persons, and the term "admission of foreigners" to connote the admission of foreigners into Turkey.
4   Law no. 7564 dated 24.07.1950, Official Gazette 24.07.1950-7564.
5   Law no. 6458 dated 04.04.2013, Official Gazette 12.04.2013-28615.

An examination of Passport Act and LFIP together reveals that a group of foreigners is considered to be acceptable to Turkey upon the fulfilment of certain conditions, and another group of foreigners is deprived from the right to enter in Turkey. A foreigner's eligibility to enter into Turkey is frequently determined following the foreigner's application to entrance points namely the border gates. Due to that, the entrance points to Turkey rise as the first issue to deal concerning the foreigners' entrance in Turkey.

2.1 Entry Points to Turkey

Under Passport Act and LFIP, entry into and exit from Turkey could be made only through the border gates (Passport Act art. 1, LFIP art. 5); entry from any point else than the border gates are illegal.

What should be understood with the term "border gate" is determined within the scope of the Implementing Regulation on The Law on Foreigners and International Protection Law[6] (The Regulation). According to The Regulation, border gates are the land, air, sea and railway border crossing points for entry into and exit from Turkey, which are physically isolated from external environment or deemed so [Regulation art. 3/1 (cc)]. Border gates are designated by the President (Passport Act art.1) from the areas in the border regions, and the entry into Turkey outside these points shall not be treated as legal (Doğan, 2017: p. 38).

Passport Act is applicable both to Turkish and foreign nationals. Therefore, under the Passport Law, Turkish nationals and foreigners were not distinguished from each other with regard to entering into the country; every individual, regardless of her/his nationality status, is obliged to enter from the border gates[7] (Passport Act art.1). On the contrary, the scope of LFIP is limited to foreigners, thus regulates only the entry of foreigners into Turkey. For this reason, in ILFA, entry into Turkey through illegal means and sanctions of illegal entry are governed exclusively for foreigners. Under ILFA foreigners that illegally enter into or exit from Turkey or, attempt to do so are sentenced to two thousand Turkish Liras of administrative fine [LFIP m. 102/1(a)].

---

6 Official Gazette: 17.07.2016-29656.
7 On the other hand, Ministry of Interior is responsible of taking the necessary measures for the implementation of this provision not to hinder the international protection claim in the border gates.

2.2 Presentation of Passport or Passport Substitutes

The first requirement that needs to be fulfilled to enter into Turkey is the presentation of a valid passport in the border gates. However, submission of a travel document that qualifies as "passport substitutes" also enables the foreigner to enter into Turkey. The documents which could be qualified as passport substitutes are laid out in domestic laws or international treaties. Besides, Ministry of Interior and Ministry of Foreign Affairs has the power to co-decide on different documents to be accepted as passport substitutes (Passport Act art. 2/3). Therefore, the passport substitutes may vary in time.

Foreigners who do not hold a valid passport or a passport substitute shall be refused to enter into Turkey (LFIP art. 7), and if the foreigner enters in Turkey via illegal ways she/he will be subjected to removal decision (LFIP art. 54/c, h).

Documents that could be presented in the border gates are as follows:

2.2.1 Passport

The passport is generally defined as a document issued by the state to which the individual is bound and indicates that there is no legal obstacle for the holder to leave or enter into the country (Çelikel, 2018: p. 82)[8]. A foreigner who desires to enter into Turkey is obliged to submit the border gate authorities a "duly issued and valid" passport (Passport Act art. 2). The term passport covers "diplomatic, service, special and official" passports, and a "duly issued and valid passport" denotes a passport that is issued by the competent authorities, is in the validity period and is not subject to fraud. Besides, the passport should be valid for at least six months as of the foreigner's arrival in Turkey.

A person who arrives in a Turkish border gate without a "duly issued and valid" passport is refused to enter into Turkey. If a foreigner enters into Turkey without a passport by any means, he/she will be sentenced to an administrative fine between a thousand and three thousand Turkish Liras[9].

2.2.2 "Reserved for Foreigners" Passport

Under Passport Act, some of the foreigners could also be given a travel document generally named as "reserved for foreigners passport". Reserved for

---

8 Compare with Çiçekli, 2017: p. 71.
9 The abrogated 32. Article of Passport Act required the deportation of the foreigners who entered into Turkey without a passport. This provision is repealed by LFIP art. 124/1.

foreigners passport could be given to the stateless people[10], and foreigners who are deemed as *de facto* stateless due to their irregular nationality (PK m. 18, LFIP m. 51/d). Reserved for foreigners passport are given to provide such foreigners' travel abroad and prevent their deportation[11] (Çelikel, 2018: p.83). However, to avoid the claims in that vein, Passport Act lays out with a clear language that reserved for foreigners passport do not provide any right to acquire Turkish nationality.

In addition, refugees and foreigners under secondary protection are given a travel document which could also be evaluated under the scope of reserved for foreigners passport provisions (LFIP m. 84/2).

2.2.3 Transires and Similar Documents

Transires[12], administrative letters and border crossing documents are the travel documents given to the Turkish Nationals to substitute the passport in cross border travels between Turkey and the countries with a border to Turkey. These documents are given to facilitate the cross border passage of Turkish nationals who live in the border regions (Çelikel, 2018: p. 83; Çiçekli, 2016: p. 76). The form and the validity period of these documents as well as to whom the documents could be given are determined by the Ministry of Interior Affairs and Ministry of Foreign Affairs under the international treaties that Turkish Republic is a party to.

---

10 LFIP names the travel document given to such foreigners as "Stateless Person Identification Document" (LFIP art. 50/1). The Document also entitles such foreigners the right to legally reside in Turkey (LFIP art. 50/1).
11 The Passport Act sets of two different types of reserved for foreigners passport:

A) The passport valid only for one entrance into Turkey or one exit out of Turkey. These passports expire in the date of entrance in Turkey when they are given for entrance and upon the arrival in the country that is indicated in the annotations when they are given for exit.
B) Passport given for one exit one entry. These passports are given for a minimum period of three months time determined under the discretion of Ministry of Interior (PK m. 18/A, B).

12 Transires are given and issued under the international treaties concluded between Turkey and its border countries.

2.2.4 Documents Given to the Seafarer, Railway and Flight Officials and Crew

"Seafarer's identity card" is a duly issued travel document with a photo given to the Turkish crew of Turkish ships which sail outside of territorial waters of Turkey. Seamen's identity card could be issued by The Regional Port and Maritime Affairs Directorate[13]. "Flight crew document" is a travel document given to the officers and crew of international air transport vehicles by the competent authorities in accordance with the procedure and sample specified in the agreements. "Railway Crew Identity Card" is a travel document with a photo given by the competent authorities to the officers and crew of international rail transport vehicles, in accordance with the procedure and sample specified in the agreements. Turkish citizens who submit these documents may enter from the border gates without having to present their passports.

The foreign vessels, air and international railway transport vehicles officers and crew may enter into and exit out of Turkish territorial waters and port cities on the basis of "reciprocity" with their duly issued documents (Passport Act art. 20/5)[14]. The Passport Act seeks for reciprocity for the entry of these foreigners; however, it does not make any distinction for the reciprocity to be provided by laws, international agreements or *de facto*. In that case, in admission of a foreigner who holds these documents, existence of reciprocity between Turkish Republic and the State that the foreigner is a national of is sufficient; it does not matter if the reciprocity is provided by international agreements, by the laws or by the implementations.

2.2.5 Refugee Travel Document

The documents to be presented by the refugees in entrance into Turkey is regulated under Article 28 of 1951 Convention Relating to the Status of Refugees[15]

---

13 Turkey has been a party to Seafarers' Identity Documents Convention, 1958 (No. 108) of ILO since 2003 (Official Gazette: 22.07.2003-25176). According to the Convention "Each Member for which this Convention is in force may issue a seafarer's identity document to any other seafarer either serving on board a vessel registered in its territory or registered at an employment office within its territory who applies for such a document." (Art. 2/2). For the whole text of the Convention see. https://www.ilo.org/dyn/normlex/en/f?p=NORMLEXPUB:12100:0::NO::P12100_ILO_CODE:C108

14 According to Convention No 108 "Each Member shall permit the entry into a territory for which this Convention is in force of a seafarer holding a valid seafarer's identity document, when entry is requested for temporary shore leave while the ship is in port" (Art. 6/1).

15 For the whole text of the Convention see http://www.unhcr.org/1951-refugee-convention.html.

which Turkey is a party to since 1961[16]. According to the Convention, The Contracting States issue refugees travel documents for their lawful stay in their territory, and for the purpose of travel outside their territory unless compelling reasons of national security or public order otherwise require. The form and the provisions of these documents are determined in the Annex of the Convention. Schedule to this Convention shall apply with respect to such documents. Besides, according to LFIP article 84/2 travel document requests by conditional refugees and subsidiary protection beneficiaries shall be evaluated within the scope of Article 18 of Passport Act (Çiçekli, 2016: p. 79).

2.2.6 Identity Cards

As long as it is provided by bilateral or multilateral treaties, foreigners can submit identity cards instead of passports to enter into Turkey. Likewise, in accordance with the Treaty signed between Turkey and North Cyprus Turkish Republic which entered in force in 02.09.1991[17], nationals of both Countries are capable of enter into and exit from the other party's territory with their identity cards[18]. Also, according to the Protocol concluded with Georgia, Turkish and Georgian nationals enter into each other's countries by presenting their identity card[19].

Besides, under the **European Agreement on Regulations Governing the Movement of Persons between Member States of the Council of Europe**[20] (**The Agreement**) which Turkey is a party since 1961[21], nationals of the Contracting Parties, whatever their country of residence are, may enter or leave the territory of another Party by all frontiers on presentation of their identity cards for the visits

---

16 Official Gazette 05.09.1961-10927.
17 Official Gazette: 30.07.1991-20945.
18 To see the whole list visit http://www.mfa.gov.tr/countries-whose-citizens-are-allowed-to-enter-turkey-with-their-national-id_s.en.mfa.
19 Official Gazette: 22.10.2011-28092. Protocol "between the governments of Georgia and the Republic of Turkey on changes and amendments to the visa agreement of April 4, 1996" that took effect on 10 December, 2011 provides visa-free movement for Georgian citizens on the territory of Turkey during 90 days (m. 1/3). For further details see: http://www.mfa.gov.ge/MainNav/ConsularInformation/VisaInfoGeorgian/%E1%83%97%E1%83%A3%E1%83%A0%E1%83%A5%E1%83%94%E1%83%97%E1%83%98%E1%83%A1-%E1%83%A0%E1%83%94%E1%83%A1%E1%83%9E%E1%83%A3%E1%83%91%E1%83%9A%E1%83%98%E1%83%99%E1%83%90.aspx?lang=en-US.
20 For the whole text of the Agreement see: https://www.coe.int/en/web/conventions/full-list/-/conventions/rms/0900001680064588.
21 Official Gazette: 01.12.1961-10972.

of less than three months' duration or whenever the territory of another Party is not entered for the purpose of pursuing a gainful activity (**The Agreement art. 1, 2 and 3)**[22],[23].

2.3 Visa Obligation

2.3.1 *The Usual Way to Get a Visa*

2.3.1.1 *Types of Visas*

The foreigner[24] who wishes to enter into Turkey has the obligation to take visa besides presenting their passport[25]. The term visa is usually engaged to refer to a "permission"; however, the definitions brought in the doctrine vary. According to a generally accepted definition, visa is a permission taken from the official authorities of a country to enter into its territory (Çelikel, 2018: p. 87; Çiçekli, 2016: p. 80). Göğer, on the other hand, defines visa as an administrative transaction which is subject to charge that indicates the passport is in accordance with the laws of the country of issue, the passport is not subjected to fraud and enables the foreigner to enter into and exit from the country (Göğer, 1973: p. 153). With his definition Göğer emphasizes different functions of visa.

LFIP defines visa as a permission that entitles a stay up to a maximum of 90 days in Turkey or to transit through Turkey [LFIP art. 3/1(t)]. Accordingly, foreigners who want to travel transit through Turkey or who wish to stay in Turkey for a maximum of 90 within 180 days are required to obtain a visa. Visa does not enable the foreigner either to stay longer than 90 days in Turkey or to work in Turkey. A foreigner who wishes to stay in Turkey more than 90 days is obliged to get a residence permit, and the foreigner who wishes to work in Turkey is obliged to get a work permit.

---

22 In accordance with the Agreement nationals of Germany, Belgium, Holland, Spain, Switzerland, Luxembourg, Malta, Greece can enter into Turkey with their identification cards. However, due to the reservation in the Annex of the Agreement Turkish nationals do not have the right to enter into the mentioned countries (Çiçekli, 2016: p. 78).
23 For the full list of the foreigners who can enter into Turkey with their identification cards see: http://www.mfa.gov.tr/countries-whose-citizens-are-allowed-to-enter-turkey-with-their-national-id_s.en.mfa.
24 Under Turkish law visa is required only for the foreigners. Visa obligation cannot be imposed for the nationals (Constitution art. 23/6).
25 For a detailed examination of the issue see: Teksoy, 2013.

There are different types of visa. The types of visas and the modes of visa applications are regulated in The Regulation. The Regulation lays out the types of visas according to their purpose of stay as tourist visa[26], transit visa[27], airport transit visa[28], education visa[29], work visa[30], official duty visa[31] and other visas[32]. Thus, the visa also functions as a determinant for the foreigners' purpose of entry into Turkey.

2.3.1.2 Visa Application Procedures

Foreigners wishing to stay in Turkey for up to 90 days are obliged to obtain a visa that indicates the purpose of their visit from the consulates of the Republic of Turkey in their country of nationality or legal stay [LFIP art. 3/1(t)]. The foreigner who applies to obtain a visa is requested to fill a visa application form. Visa application form shall be filled separately for each foreigner (The Regulation art. 12/2). The foreigner is obligated to present information and documents, which may be requested based on visa types, to the competent authority during the application. Applications of foreigners who cannot collect the requested documents within the identified period of time, cannot be subjected to evaluation (The Regulation art. 12/3). The assessment of applications lodged with consulates shall

---

26 Tourist visa is issued to foreigners who wish to come to Turkey for the purposes of touristic or official visits, business meetings, conferences, seminars, meetings, festivals, fairs, exhibitions, sporting events, cultural and artistic events [The Regulation art. 11/1(a)].
27 Transit visa is issued to foreigners, who wish to enter Turkey through any border gate and cross through Turkey within a determined period of time [The Regulation art. 11/1(b)].
28 Airport transit visa is issued to foreigners, who only wish to cross through air border gates without entering Turkey [The Regulation art.11/1(c), art. 16/4].
29 Education visa is issued to foreigners, who wish to arrive in Turkey for the purpose of education, training, internships, courses, student exchange programs, Turkish language courses [The Regulation art.11/1(ç)].
30 Work visa is issued to foreigners within the scope of Article 55 of the Implementing Regulation of the Law on Work Permits of Foreigners (Official Gazette 29/08/2003, No. 25214), and foreigners, who are not within this scope and wish to come in order to work [The Regulation art. 11/1(d)].
31 Official duty visa is issued to foreigners, who are appointed to an official post or appointed as a diplomatic courier [The Regulation art. 11/1(e)].
32 Other visas are issued to foreigners, who are not within the scope of the above mentioned visa types and wish to come to Turkey for the purposes including archeological excavation, research, movie or documentary shooting, treatment, accompany, family unification, humanitarian assistance, transportation [The Regulation art. 11/1(f)].

be determined within 90 days (LFIP art. 11/1,4). Visas are issued by the consulates and, in exceptional cases[33], by the governorates in charge of the respective border gates (LFIP art. 11/4; LFIP art. 13/2).

A relatively new progress is the opportunity for the nationals of the countries that are determined by the President is the Electronic Visa[34] (e-Visa) application (The Regulation art. 12/1). The e-Visa Application System was launched on 17 April 2013 by the Ministry of Foreign Affairs which allows foreigners travelling to Turkey to apply and obtain their e-Visas online[35]. However, e-visa is only valid when the purpose of travel is either tourism or commerce. For other purposes, such as work and study, visas are given by Turkish Embassies or Consulates.

On the other hand, under LFIP art. 18/1(a) the President is authorized to "enter into agreements determining the passport and visa procedures; and under circumstances when considered necessary, unilaterally waive the visa requirement for nationals of certain states; facilitate visa procedures, including exemption from visa fee; and, determine the duration of visas". Thus, LFIP entitles the President to facilitate the visa procedures for the nationals of desired States. In this context, nationals of a group of states are able to obtain visa during their entrance into Turkey[36]. This type of visa is generally named as "stamp visa" or "visa on arrival". Currently, nationals of over 80 countries can benefit the opportunity to get visa on arrival[37].

Besides, the President is also entitled to introduce terms and conditions for the use of passports belonging to foreigners with regard to entry into or stay in or exit from Turkey, in case of war or other extraordinary circumstances to cover a region of or the entire country [LFIP art. 18/1(a)], and take all measures setting specific conditions or restrictions regarding entry of foreigners into Turkey [LFIP art. 18/1(a)]. In the existence of such conditions, and measures, procedures concerning visa should be evaluated in this context.

---

33   The exceptional cases are set out under The Regulation. For a detailed information see, section below.
34   For further info on e-visa application see https://www.evisa.gov.tr/en/tour/.
35   For the full list of countries whose nationals are eligible for e-visa application see https://www.evisa.gov.tr/en/info/.
36   The national of these states can also benefit from e-Visa application. By this application the mentioned foreigners can obtain visa via internet from the web site "Republic of Turkey e-Visa Application System". See https://www.evisa.gov.tr/en.
37   For the list of the States whose nationals can benefit from this convenience see Ekşi, 2018: p. 83–85.

2.3.2 Exceptional Visa

Besides the mentioned visa types and visa conveniences, LFIP art. 13 sets out an exceptional visa (Çelikel, 2018: p. 89) that is named as "border visa"[38]. Border visas are issued in the border gates due to necessities. Therefore "border visa" differs from the visa conveniences that is accorded by the President since it could be issued only out of necessities (Çelikel, 2018: p. 89; Ekşi, 2018: p. 82).

The situations that could be qualified as "necessities" are set out in The Regulation According to Article 15 of The Regulation, border visa could be given,

- In case of the absence of a consulate in the country, where foreigner legally resides, or visa proceedings cannot be conducted through consulates.
- In case of compelling reasons including disease, death or accident of the foreigner or his/her spouse and close relatives.
- Upon the notification of risks by the related units in terms of health condition of the foreigner in case of his/her removal.
- Upon arrival of a foreigner to participate in national and international scientific, economic, cultural and commercial events.
- Upon arising of the obligation that commercial vessel crews, which arrive at sea ports, to continue moving towards their own country or another country or to join another vessel at the same port.
- Upon notification by public institutions and organizations that a visa needs to be issued to a foreigner at the border.

Under these cases, excluding the foreigners whose entry into Turkey would not be permitted, governorates may issue visas to foreigners who arrive at border gates without obtaining a visa on condition that they document their conditions mentioned above and depart from Turkey in due time within the framework of the exceptional circumstances (Regulation art. 15/1). Such visas authorize stay in Turkey for a maximum of 15 days, unless a different duration is determined by the President.

2.3.3 Foreigners Who Shall Be Refused to Take Visa

LFIP art. 15/1 determines the foreigners who cannot be granted visa, by listing the reasons and the persons who are not eligible to enter into Turkey. While setting

---

38 According to Çelikel, Airside transit visas that are set out in Article 14 of LFIP is another type of exceptional visa. Airside transit visas could be requested from foreigners who are transiting through Turkey (LFIP art. 14). Airside transit visas could be issued by the consulates, to be used no later than six months (LFIP art. 14/1).

out the foreigners that cannot be granted a visa LFIP also indirectly shows the conditions that a foreigner is obliged to fulfill for the visa application. As a principle the foreigners listed in LFIP art. 15/1 cannot be granted visa and thus are not allowed to enter into Turkey. However, LFIP art 15/2 provides an exception to the first paragraph. According to LFIP art. 15/2, if it is deemed to be of interest to issue a visa to such a foreigner who falls within the scope of this article, a visa may be granted subject to the Ministry of Interior's approval.

The foreigners whose visa application shall be refused are as follows:

2.3.3.1 A Foreigner Whose Passport or Travel Document Is Not Valid At Least Sixty Days Beyond the Expiry Date of the Visa Requested (LFIP art. 15/1)

The first condition of being granted a Turkish visa is to hold a passport or a travel document which is 60 days beyond the expiry date of the visa requested. A foreigner who holds a passport or a travel document that does not fulfill this condition cannot be granted visa [LFIP art. 15/1(a)].

2.3.3.2 Foreigners Who Are Banned from Entering Turkey

Under LFIP art. 9 The Directorate General, when necessary and upon consultation with the relevant government departments and institutions, may impose an entry ban against foreigners whose entry into Turkey is objectionable for public order, public security or public health reasons (LFIP art. 9/1). Also, The Directorate General or governorates imposes an entry ban for foreigners who are deported from Turkey [LFIP art. 9/2, LFIP art. 15/1(b)][39].

2.3.3.3 Foreigners Who Are Considered Undesirable for Reasons of Public Order or Public Security

Foreigners who are considered undesirable due to "public order" or "public security" reasons cannot be granted visa. However, application of this provision may cause to problem since the term "public order" and "public safety" are ambiguous [LFIP art. 15/1(c)]. In this case, determination of "undesirability" of a foreigner with regard to public order or public security remains as a matter left to the discretion of the administrative bodies (Teksoy, 2013: p. 882). Even though it is usual for the administrative authorities to be given powers of discretion, it is also necessary to limit the administrative authority with the obligation to rely on concrete facts and conduct an adequate examination. Decision on the rejection

---

39 This subject will be evaluated in the Third Chapter in detail.

of the alien's application should be taken in an objective and not arbitrary fashion (Teksoy, 2013: p. 882).

2.3.3.4 The Foreigners Who Are Identified to Have a Disease Posing Public Health Threat

According to LFIP, a foreigner who is eligible for visa should be free of any diseases that would threat general health [LFIP art. 15/1(ç)]. Under LFIP not "any kind of diseases" but only the ones who threaten the general public health is undesirable. In identification of diseases qualified as a threat to public health, Law No. 1593 on Public Health[40] is applicable and in determining the whether the disease is a contagious or contagious parasite disease and is potentially epidemic (Regulation art. 17/2).

2.3.3.5 Foreigners Who Are Suspects of or, Are Convicted of a Crime That Are Subject to Extradition Pursuant to Agreements to Which the Republic of Turkey Is a Party to

Foreigners who are suspects of or, are convicted of a crime that are subject to extradition pursuant to agreements to which the Republic of Turkey is a party to cannot be granted visa (LFIP art. 15/1-d). From the clear language of LFIP, it can be concluded that it is not necessary for the foreigner to be convicted of the relevant offense; the fact that her/his being accused of the relevant crime is sufficient for refusal of visa application.

2.3.3.6 Are Not Covered with a Valid Medical Insurance for the Duration of Their Stay

In order for a foreigner to be regarded as eligible, she/he should have a valid medical insurance that would cover the duration of her/his stay in Turkey [LFIP art. 15/1(e)]. The foreigners who do not have a valid health insurance which would meet the health expenses cannot be granted a visa. The intention of this provision is to protect the Turkish health and social security system in financial means and on the other hand prevent grievances that foreigners may confront. Nevertheless, an exemption to this condition is set out under LFIP art. 13. According to this provision, the medical insurance requirement may be waived for humanitarian reasons for persons who are issued a visa at the border (LFIP art. 13/3).

---

40 Official Gazette: 06.05.1930-1489.

2.3.3.7 The Foreigners Who Fail to Supply Proof of the Reason for Their Purpose of Entry into, Transit from or Stay in Turkey

The foreigners who apply for visa are obliged to indicate their purpose of entry and stay, and ground it on legal reasons [LFIP art. 15/1(f)]. Likewise the competent authorities determine the foreigner's purpose of entry, and stay via the visa types granted to the foreigner. The purpose could be touristic, commercial, academic, cultural or professional as well work, education or residence. Unless the foreigner proves that her/his entry into, stay in or transit from Turkey bases on reasonable and legal grounds, the foreigner is refused to be granted visa.

2.3.3.8 Foreigners Who Do Not Possess Sufficient and Sustainable Resources for the Duration of Their Stay

The foreigners who do not possess sufficient and sustainable financial resources cannot be granted visa [LFIP art. 15/1(g)]. The term "sufficient" is not explained neither in LFIP nor in The Regulation, thus determining the sufficiency of the financial sources are left to the discretion of administrative authorities. In determining the sufficiency of the financial resources, the current satiation of the foreigner, and the duration and purpose of her/his stay should be taken in consideration (Teksoy, 2013: p. 885).

2.3.3.9 Foreigners Who Refuse to Pay Receivables, Originating from Overstaying the Duration of Visa or a Previous Residence Permit Duration or Some Other Debts or Fines

Under LFIP, the foreigner needs to be free of some fine and debts or she/he acknowledges to pay them [LFIP art. 15/1(ğ)]. Even though the wording of LFIP suggests that the "acknowledgement of the payment of these debts and fines is sufficient", it would be more appropriate to interpret this provision as the actual payment of this debts. (Teksoy, 2013: p. 885).

The receivables that are required by the law to be paid are:

- Receivables originating from overstaying the duration of visa
- Receivables originating from previous residence permit duration
- Receivables that should be enforced and collected pursuant to the Law on the Procedure of Collection of Public Receivables No. 6183 of 21.07.1953[41]
- Debts and fines enforced pursuant to the Turkish Penal Code No. 5237 of 26.09.2004[42]

---

41 Official Gazette: 28.07.1953-8469.
42 Official Gazette: 12.10.2004-25611.

2.3.4 Consequences of Getting a Visa

The foreigner who wishes to enter into Turkey has the obligation to have a valid and dully issued visa (LFIP art. 11/2). The absence of the visa hinders foreigner entrance into Turkey. Thus, visa functions as an administrative transaction that provides the foreigner's entrance into Turkey and shows her/his purpose of visit to Turkey. However, possessing the visa does not confer the foreigner an absolute right of entry (LFIP art.11/3). The fact that a foreigner had obtained a visa shall not prevent the implementation of provisions of the Law related to foreigners, whose entry into Turkey would not be permitted during entry into the country. Foreigner shall not be permitted to enter the country, if she/he has been identified to be among persons, whose entry into Turkey would not be permitted.

2.3.5 Cancellation of Visa

The visas that could be granted to a foreigner can be subject to cancellation. Visa can be cancelled by the issuing authorities or the governorates in cases when/where:

- It is determined that the visa is exploited for fraudulent purposes [LFIP art. 16/1(a)].
- There is erasure, scraping or alteration detected on the visa sticker [LFIP art. 16/1(b)].
- The visa holder is banned to enter into Turkey [LFIP art. 16/1(c)].
- There is strong doubt as to the foreigner may commit a crime [LFIP art. 16/1(d)].
- The passport or travel document is false or has expired [LFIP art. 16/1(e)].
- The visa or the visa exemption is used outside of its purpose [LFIP art. 16/1(f)].
- The circumstances or documents on the grounds of which the visa was issued are determined to be not valid [LFIP art. 16/1(g)].

However, according to the system adopted by LFIP, the reasons for the cancellation of visa do not result in direct invalidity of the visa; the visa must be cancelled by an administrative decision (Teksoy, 2013: 885). In the event that the conditions listed in ILFA art. 16/1 have occurred, the administrative authorities have to give a decision of cancellation. The authorities do not have discretionary authority in this respect. They will use the discretionary authority only to evaluate if the certain incidents match with the conditions specified in ILFA art. 16/1 (Teksoy, 2013: p. 885).

Besides, the reasons specified in LFIP art. 16/1, ILFA art. 16/2 regulates another situation which requires the visa to be cancelled. Accordingly, the visa is cancelled in case a removal decision is issued for the foreigner within the duration of the visa (LFIP art. 16/2).

Visa can be cancelled by the issuing authorities or the governorates (LFIP art. 16/1). The processes related to the cancellation of the visa shall be notified to the visa applicant (LFIP art. 17). The decision of cancellation can enforceable upon the notification.

2.3.6 Visa Exemption

LFIP defines visa exemption as the regulation waiving the visa requirement [LFIP art. 3/1(u)]. With a more detailed explanation, visa exemption could be defined as the ability of a foreigner to enter into Turkey without a visa and stay from a period of time provided by the visa exemption.

Visa exemption could be provided by the law or by international treaties.

2.3.6.1 Visa Exemption Provided by Law

The foreigners eligible for visa exemption are listed under LFIP. Under LFIP art. 12 some groups of foreigners are granted to absolute exemption, where some other groups of foreigners' exemption are left to the discretion of the administrative authorities.

2.3.6.1.1 Foreigners with Absolute Visa Exemption

2.3.6.1.1.1 Foreigners Exempt from Visa Obligation with a Presidential Decree

This kind of exemption derives from the unilateral decrees of the President [LFIP art. 12/1(a)]. It is possible for the President to grant visa exemption to nationals of certain countries by taking into account international relations or by making political decisions that may be taken for economic reasons.

2.3.6.1.1.2 Holders of a Residence or a Work Permit Valid on the Date of Entry into Turkey

####### 2.3.6.1.1.2.1 Holders of a Residence Permit

LFIP defines residence permit as the permit issued for the purpose of staying in Turkey. Foreigners who would stay in Turkey beyond the duration of a visa, and in any case longer than 90 days should obtain a residence permit [LFIP art. 3/1(j)]. The holders of residence permit are not sought to have a visa [LFIP art. 12/1(b)]. LFIP specifies six types of different residence permit LFIP art. 12/1(b)[43]. However,

---

43 The types of residence permit are short-term residence permit, family residence permit, student residence permit, long-term residence permit, humanitarian residence permit and victim of human trafficking residence permit (LFIP art. 30).

LFIP art. 12/1(a) does not make any distinction with regard to entrance into Turkey. In this case, holders of any kind of residence permit are exempted from visa in entering into Turkey.

One issue that needs to be addressed in this regard is whether the residence permit exemption or the documents that substitute the residence permit could be admissible as to grant visa exemption. Regarding the issue Teksoy, timely states as "the liberty recognized for the more and more important, should be recognized as *argumentum a fortiori* for less and less important. Moreover, it is not possible for freedom of residence to be used before the entry into the country takes place. From the point of view of the fact that the entrance can be subjected to visa while the residence is released seems meaningless as well." (Teksoy, 2013: p. 869)

In that case, the foreigners who are exempt from visa due to residence permit or residence permit exemption are as follows:

- Valid residence permit holders [LFIP art. 12/1(b)].
- Foreigners who fall within the scope of Article 28 of Law № 5901.
- Holders of stateless person identification card [LFIP art. 51/1(a)].
- Foreigners International Protection Applicant Identity Document (LFIP art. 76/d).
- Identity document for international protection beneficiaries (LFIP art. 20/1).
- Members of the diplomatic and consular missions in Turkey.
- Family members of diplomatic and consular officers, provided they are notified to the Ministry of Foreign Affairs.
- Members of the representations of international organizations in Turkey whose status has been determined by virtue of agreements.
- Foreigners who are exempt from a residence permit by virtue of international agreements which Turkey is a party to.
- Holders of migrant document under Resettlement Law.
- Turquois Card[44] holders (International Labor Force Act art. 11/3, 4).
- Holders of work permit exemption [LFIP art. 12/1(b), The Regulation art. 13/1(b)].
- Holders of reserved for foreigners passport, namely stateless and *de facto* stateless foreigners and subsidiary protection beneficiaries (Passport Act art. 18, LFIP art. 51/d), conditional refugees and subsidiary protection beneficiaries (LFIP art. 84/2).

---

44 Turquois Card is the document which grants indefinite work permit to the foreigner in Turkey and residence permit to his/her spouse and dependent children in accordance with the provisions of the legislation (ILFA art. 3/ğ). "Turquois Card" denotes to a facilitated work permit system for "high-qualified foreigners" who would contribute to Turkey in different means.

2.3.6.1.1.2.2 Holders of a Work Permit

Foreigners could work in Turkey provided that they hold a work permit or work permit exemption. The provisions of work permit and work permit exemption are regulated under International Labor Force Act (ILFA). ILFA defines work permit as "The permit that is issued by Ministry of Interior and grants residence permit to the foreigner until its expiration" [ILFA art. 3/1(c)]. Therefore, work permit also grants the right to reside, besides the right to work. Under ILFA work permits are subjected to a separation according to their validity period or their subject. There are five types of work permits that are regulated in ILFA[45]. Since LFIP does not make any distinction between the work permit types with regard to entry into Turkey, foreigners with any kind of work permit can enter in Turkey.

With regard to the work permit exemption, like for the residence permit exemption the question whether the work permit exemption or the documents that substitute the work permit could be admissible as to grant visa exemption emerges. In compatible with the evaluation made for the residence permit exemption, the question should be answered as "yes" and work permit exemption should also be accepted as visa exemption.

However, the eligible foreigners for work permit are not defined in a separate article in ILFA (Çelikel, 2018: p. 200; Yılmaz, 2017: p. 97) since the detailed regulation on the determination of the foreigners eligible for work permit are left to "The Regulation on the Application of International Labor Force Act" which is yet not issued [ILFA art. 25/1(b)] (Çelikel, 2018: p. 200; Yılmaz, 2017: p. 97). Nevertheless, two different articles of ILFA, ILFA art. 13/7 and ILFA art. 14, mentions about the foreigners that are eligible for work permit exemption.

According to ILFA art. 13/7,

- Executive board members of the joint stock companies which established in Turkey under Turkish Commercial Code Law no 6102 of 13.01.2011[46] (Turkish Commercial Code) that are not domiciled in Turkey.
- Partners of companies which are established in Turkey under Turkish Commercial Code that are not directors.
- Cross-border service providers whose activities carried out in Turkey does not exceed 90 days within 180 days shall be evaluated within the scope of work permit exemption.

---

45 The types of work permits are temporary work permit (ILFA art. 10/1), indefinite work permit (ILFA art. 10/3), independent work permit (ILFA art. 10/5), Turquois Card (ILFA art. 15), exceptional work permit (ILFA art. 15).
46 Official Gazette: 14.02.2011- 27846.

Under ILFA art. 14, in the diplomatic and consular representations of foreign countries in Turkey, foreigners who work in schools, cultural institution and religious institutions which are operated as organizational units are able to work with work permit exemption. Also, for the foreigners who work as diplomatic staff, consulate officer, administrative and technical staff and consulate attendant in the diplomatic and consular representations of foreign countries in Turkey and the foreigners who work in international organizations in Turkey as international officer and administrative and technical staff, and the foreigners that work in their private service are able to work with work permit exemption.

c) **Holders of a Valid "Reserved For Foreigners" Passport**

Holders of "reserved passport foreigners"[47] can enter into Turkey without any visa obligation [LFIP art. 12/1(b)].

ç) **Foreigners within the Scope of Article 28 of the Turkish Nationality Act № 5901 of 29/05/2009**[48]

Turkish nationals by birth who lost Turkish nationality by obtaining renunciation permit and their lineal kins down to the third degree do not need to get a visa to enter into Turkey. According to Turkish Nationality Act (TNA) art. 28, such foreigners hold the right to benefit from the same rights accorded to Turkish nationals except for the military service obligation, the right to vote and be elected, the right to be employed in public services (TNA art. 28/2), right to import exempted vehicles and household goods (TNA art. 28/2), provided that the provisions concerning national security and public order are reserved (TNA art. 28/1). Therefore, since these foreigners benefit from the rights akin to a Turkish national's rights, they are not obliged to obtain visa to enter into Turkey.

2.3.6.2 Visa Exemption Provided by the International Agreements

Art. 16 of The Constitution states that "The fundamental freedom and rights in respect to aliens may be restricted by law compatible with international law" (The Constitution Art. 16) and enables restricting the fundamental freedom of foreigners provided that these restrictions are made by law and in compliance with the international law (Ekşi, 2018: p. 229; Gözler, 2017: p. 468). Therefore foreigners' right to enter into Turkey cannot be restricted in a fashion contrary to international law.

---

47 For a detailed examination of the issue see "Section 1.2.2".
48 Official Gazette: 12.06.2009-27256.

International law recognizes an exhaustive discretion for the states in restricting the entrance of the foreigners into their territory. However, when a state becomes a party to an international treaty regarding the issue, since the treaty is also an international law tool, the state has to obey with the Treaty. Turkey's obligation to honor these treaties and adopt regulations in compliance with the international Treaties with regard to the rights of foreigners primarily base on this provision. Besides, LFIP art. 2/2 emphasizes the implementation of the provisions of LFIP without prejudice to international agreements (Teksoy, 2013: p. 864).

Turkey is a party to a large number of bilateral visa exemption treaties[49]. In these treaties, the visa exemption for the nationals of the state party is recognized for 30–90 days, and for non-profitable travels (Çelikel, 2018: p. 92). Besides, Turkey is a party to multilateral treaties and conventions that are providing visa exemption. The most preeminent treaty among those is the **European Agreement on Regulations Governing the Movement of Persons between Member States of the Council of Europe which Turkey became a party to in 1961.** Under The Agreement nationals of a contracting party can enter into the territory of another contracting party as long as their visits do not exceed three months' duration or do not include the purpose of pursuing a gainful activity (Art. 1/2, 3). Thus, under the agreement nationals of Council of Europe member states exempt from visa can log in Turkey[50] (Tekinalp, 1997: p. 495).

2.4 Control in the Border Gates

A foreigner logging in the border gates of Turkey is primarily subjected to a control by the border authorities regarding the authenticity and validity of the presented document. The foreigner can enter into Turkey, only if she/he is found to be appropriate for the entrance after the examination of the documents.

There are two types of control that can be conducted in the border gates:

---

49   For a list of the bilateral visa exemption agreement that Turkey is a party to see http://ua.mfa.gov.tr/ (as of 05.06.2018).
50   Article 7 of The Agreement enables the Part States to suspend or to delay the entry in force of the Agreement on grounds relating to public order, security or public health, in respect of all or some of the other Parties (The Agreement art. 7/1). Basing on this provision, majority of the States party to The Agreement suspended visa exemption for Turkish nationals, on the basis of increasing arrival of unregistered Turkish workers, and insubstantial asylum claims (Özkan, 2017: p. 412).

2.3.1 Minimum Check

The foreigner arriving at the border gate is primarily passed through the border check, which can be described as the first step of the control at border gates (Asar, 2018: 42). Border check begins when the foreigner comes to the passport officer and submits her/his documents. At this stage, the border officer takes the passports or the substituting travel documents, determines the document type and accordingly determines the visa obligation of the document holder. In this context, it shall also be checked whether the documents presented by the foreigner are valid and duly issued, whether the document belongs to the person submitting the document, and whether the foreigner is one of the foreigners who shall be refused to enter into Turkey. If any discrepancy in the documents presented by the foreigner is not recognized, the passport or the travel document substituting the passport is given back and the foreigner's passage from the border is provided.

2.3.2 Comprehensive Check

Comprehensive control is introduced in the Turkish legislation by LFIP, even though it was a *de facto* implemented even before the adaptation of LFIP (Asar, 2018: 40). Comprehensive check is regulated under LFIP art. 6/5. Under this provision if a negativity is detected during the minimum control, the foreigner could be subjected to comprehensive control. Comprehensive control proceedings could be conducted at a place near passport control point and a separate location, considering the security and personal privacy of the foreigner (The Regulation art. 6/2).

The details of the proceedings of comprehensive control is specified in The Regulation (Regulation art. 6/3). These proceedings generally cover:

- Control and examination of travel documents.
- Review of restrictions related to the records of entry into the country.
- Whether the foreigner is subject to a proceeding by judicial and administrative authorities.
- Determination of whether she/he is wanted at international level.
- Determination of the purpose of arrival in the country.
- Clarification on how the foreigner will earn her/his livelihood in Turkey.
- Investigation of whether she/he is among foreigners, whose entry into Turkey would not be permitted and for whom visa would not be issued.
- Other matters deemed necessary in terms of public security.

Comprehensive control proceedings are not administrative detentions and should be finalized in maximum four hours. Judicial proceeding periods are not included in this period. The foreigner's consent is sought, if the four-hour period would be exceeded. During the comprehensive control period, a foreigner may return to her/his country or wait for the finalization of the proceedings related to her/his entry into the country (The Regulation art. 6/2). The foreigner's access to basic needs are ensured during the control, and the foreigner is be informed about the purpose and procedure of the control. (The Regulation art. 6/2).

Procedures and proceedings related to comprehensive controls are finalized by the governorates, to which border gates are affiliated with (The Regulation art. 6/5). In the end of the comprehensive control, if the border authorities come to a negative conclusion, the foreigner is qualified as "inadmissible" and refused to enter into Turkey; if the border authorities come to a positive conclusion, entrance of the foreigner into Turkey is provided.

3 Inadmissibility to Turkey

Under LFIP, the administrative authorities are granted the right to prevent the foreigners' entry into Turkey. Foreigners' entrance could be prevented by the qualification of the foreigner as "inadmissible" or with an entry ban decision.

3.1 Foreigners Who Shall Be Refused to Enter into Turkey

During the minimum check the foreigners who are qualified as suspicious are subjected to comprehensive check. In the end of comprehensive check foreigners who are qualified as inadmissible are prevented from entering into Turkey (LFIP art. 7/1). These foreigners are the foreigners:

- Who do not hold a passport, a travel document, a visa or, a residence or a work permit or, such documents or permits has been obtained deceptively or, such documents or permits are false [LFIP art. 7/1(a)].
- Whose passport or travel document expires 60 days prior to the expiry date of the visa, visa exemption or the residence permit [LFIP art. 7/1(b)]
- Foreigners who are refused to take visa even if they are exempted from a visa [LFIP art. 7/1(c)].

Actions taken in this regard are notified to foreigners who are refused entry (LFIP art. 7/2). This notification also includes information on how foreigners would effectively exercise their right of appeal against the decision as well as other legal rights and obligations applicable in the process.

In cases where foreigners are refused entry and into or transit to Turkey and qualified as "inadmissible passenger", they are transferred to the country they came from or to a country where they shall definitely be admitted (LFIP art. 98).

Admission of Foreign Real Persons in Turkey 167

3.2 Entry Ban to Turkey

Refusal of a foreigner's entry into Turkey can also be provided by an "entry ban". LFIP art. 9 determines the procedure of issuing an entry ban while it specifies the foreigners about whom entry ban decision could be taken. According to LFIP art. 9, the Directorate General of Migration Management (Directorate General) may impose an entry ban against foreigners whose entry into Turkey is objectionable for public order, public security or public health reasons. While taking the entry ban decision Directory General may consult to the relevant government departments and institutions when it is deemed necessary.

Another group of foreigners that could be banned from entering into Turkey are the foreigners who are deported from Turkey. Entry ban to such foreigners is imposed by The Directorate General or governorates (LFIP art 9/2).

The duration of the entry ban to Turkey cannot exceed five years. However, in cases where there is a serious public order or public security threat, this period may be extended for a maximum of an additional ten years by the Directorate General (LFIP art. 9/3). On the other hand, the entry ban to Turkey for foreigners whose visa or residence permit has expired and who has applied to the governorates to exit from Turkey before their situation is established by the competent authorities upon which a removal decision has been taken, cannot exceed one year (LFIP art. 9/4). Among those who have been invited to leave Turkey, an entry ban might not be imposed for those who leave the country within the specified period of time (LFIP art. 9/5). By these last two provisions LFIP aims to encourage voluntary exit (Çelikel, 2018: p. 97).

The entry ban shall be notified to foreigners by the competent authority at the border gate when they arrive to enter into Turkey. Whereas, foreigners who are deported from Turkey are notified by the governorates. The notification also includes information on how foreigners would effectively exercise their right of appeal against the decision as well as other legal rights and obligations applicable in the process (LFIP art. 10).

4 Conclusion

The accelerated increase in the number of people in international circulation has made the determination of the conditions of entering into the country a significant issue for each state. Turkey is not an exception in this regard. Even though it is associated with security issues time to time, the number of foreigners who come to Turkey for the purposes of trade, education, tourism, etc., increase constantly. Consequently, conditions of foreigners' entry into Turkey emerges as one of the hot topics both for the Turkish State and the foreigners who wish to come to Turkey.

The conditions of entry into Turkey for foreigners have been arranged in the Passport Law for a long period. In 2013 certain provisions of Passport Law was abrogated with the adaptation of LFIP. Currently, Passport Law and LFIP are implemented in conjunction, but the majority of the provisions concerning the entry of the foreigners are regulated in the Law on Foreigners and International Protection.

With regard to the admission of foreigners in Turkey, LFIP aims to establish a balance between three considerations: First one among these is to simplify the procedures to a reasonable extent for the foreigners who come to Turkey for the purposes of tourism, commerce, cultural activities, education, etc. and meet the entry conditions. The second one is to pursue the security and financial concerns by determining the foreigners who are inadmissible to Turkey. Finally, LFIP aims to facilitate the entry of foreigners who are deemed as beneficial to Turkey, or who have cultural or ethnic connections with Turkey. In this regard several group of foreigners such as the business people in key positions, the former Turkish nationals who relinquished Turkish nationality by permission and foreigners from Turkish ancestry are granted visa exemption.

Like many other countries Turkish Law grants a broad authority to administrative authorities in admission or refusal of the entry of foreigners. This is compatible with the global tendency as well. LFIP seems to set out objective criteria for the visa requirements while it specifies the foreigners who are inadmissible. However, terms such as "public order" and "public safety" which are engaged in testing the fulfilment of these requirements present vagueness and are open to interpretation. Even though the discretion granted to administrative authorities in this regard is understandable, the introduction of criteria to specify the features of these terms would be incisive to prevent the arbitrary use of this authority, and amplify legal security.

References

Asar, A. (2018). *Yabancılar Hukuku (Law of Aliens)*, 3. Baskı, Ankara: Seçkin Yayınları.

Aybay, R. (1979). "Bir İnsan Hakkı Olarak Uluslararası Seyahat Özgürlüğü" (Freedom to Travel as a Human Right), *TODAİE İnsan Hakları Yıllığı*, 1 (1): 80-93.

Cengiz, F. Ç. (2018). "Divergent Domestic Processes for Turkey and The European Union With Regard to Convergent Migration Policies", In D. K. Dimitrov, D. Nikoloski, R. Yılmaz (Ed.). *Proceedings of International Balkan and Near Eastern Social Sciences Congress Series* (Vol: 1, pp. 301-307), Tekirdag/Turkey, March 24-25, 2018.

Çelikel, A. (2018). *Yabancılar Hukuku (Law of Aliens*, 23. Baskı, İstanbul: Beta Basım A.Ş.

Çiçekli, B. (2016). *Yabancılar ve Mülteci Hukuku (Law of Aliens and Refugees)*, 6. Baskı, Ankara: Seçkin Yayınları.

Doğan, V. (2016). Türk Yabancılar Hukuku *(Turkish Law of Aliens)*, Ankara: Savaş Yayınevi.

Ekşi, N. (2018). *Yabancılar ve Uluslararası Koruma Hukuku (Law of Aliens and International Protection)*, 4. Baskı, İstanbul: Beta Basım A.Ş.

Göğer, E. (1973). *Pasaport Hukuku*, 1. Baskı, Ankara: Ankara Hukuk Fakültesi Yayınları No. 325.

Gözler, K/Kaplan, G (2017). *İdare Hukukuna Giriş (Introduction to Administrative Law)*, 23. Baskı, Bursa: Ekin Basın Yayın Dağıtım.

Özkan, I. (2007). Türk Vatandaşlarının Avrupa Ülkelerine Giriş Hakkı ve Vize Sorunu (Entry of Turkish Nationals to Europe Countries and the Issue of Visa), *Dokuz Eylül Hukuk Fakültesi Dergisi*, Özel Sayı (9): 409–446.

Seviğ, V. R. (1977). Yabancı Gerçek Kişilerin Giriş, Çıkış ve Seyahatleri, *İstanbul Hukuk Fakültesi Mecmuası*, 1–4 (43): 367–382.

Tekinalp, G. (1998). Avrupa Birliğinde İkamet ve Yerleşmeyle İlgili Düzenlemeler ve Türkiye, *Milletlerarası Hukuk Bülteni*, 17–18 (1): 493–516.

Teksoy, B. (2013). 6458 Sayılı Yabancılar ve Uluslararası Koruma Kanununa Göre Yabancıların Vize Alma Zorunluluğu, (The Obligation of Aliens to Get Visa under the Law on Foreigners and International Protection[51] no. 645) *Ankara Üniversitesi Hukuk Fakültesi Dergisi*, 62 (3): 855–906.

Yılmaz, M. (2017), Türk Yabancılar Hukukunda Yabancıların Çalışma İzni (Working Permit in Turkish Law of Aliens), İstanbul: Oniki Levha Publication.

---

[51] Law no. 6458 dated 04.04.2013, Official Gazette 12.04.2013-28615.

Osman Erdal Şahin

Planning and Spatial Regulation in İstanbul from Tanzimat to the Republican Era

1 Introduction

Alterations in urban spaces and development activities are closely related with intellectual and political transformation of the society. Development activities during the last century of the Ottoman Empire left salient marks on urban spaces. It is also possible to consider the urban development of İstanbul from Ottoman to Republican times as a "conflict of political powers" (İgüs and İsmailoğlu, 2016: 120). Having hosted the royal residence during the Ottoman era, İstanbul never lost the advantages of this characteristic even in the Republican period. With that in consideration, solutions for problems of the province as well as its development have always attracted the states' attention. In that respect, styles and images projected for İstanbul have also had an influence on solutions.

Igniting the sparks of countless urban plans and solutions of critical importance, the Tanzimat era coincides with the institutionalization and large-scale implementations of urban planning in industrialized European states. Senior bureaucrats of the Ottoman Empire were influenced by the planning activities implemented in Western cities. That also reflected the impact of industrialization on urban spaces. Meanwhile, urban development and planning became institutionalized in the Western world, and Ottoman administrators proceeded to imitate them. Western administration and planning models began to be imitated in the aftermath of fire outbreaks due to the wooden texture of İstanbul (Tekeli, 2013: 39). Urban planning and development of İstanbul, therefore, reflected the political mentality of the period from the Ottoman to the Republican times. Public spaces of İstanbul during the final century of the Empire were regulated with the intent of attaining symbolic plans and goals of the government in power.

2 From Moltke's Plan Towards Newer Planning Operations

The first scale map of İstanbul, capital of the Ottoman Empire for many years, was drawn by François Kauffer in 1786. Depicting Üsküdar, Galata, the historical peninsula, the Golden Horn as well as the Bosphorus, this map constituted the first example of development and planning activities in İstanbul. As the first

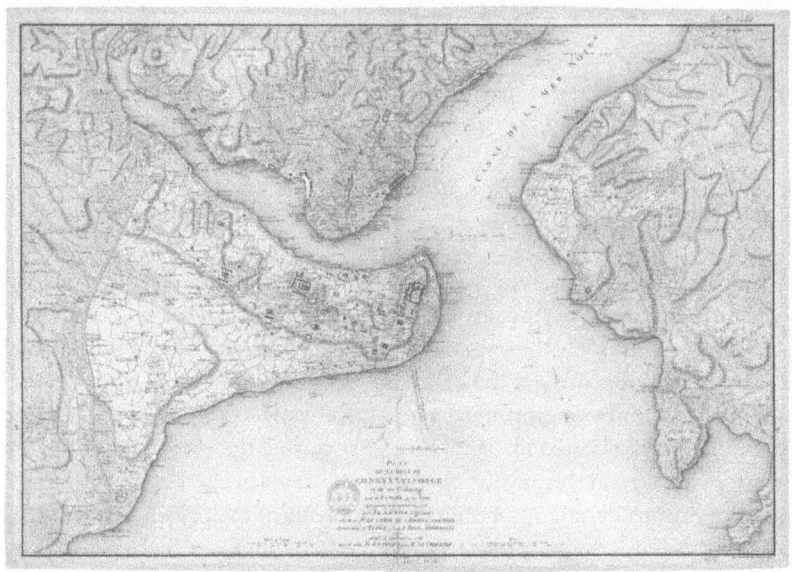

Fig. 1: François Kauffer's Map of İstanbul. Source: (https://İstanbul-constantinople.culturalspot.org)

scientific map of the city, this map had always had an important position in the drawing of further detailed maps and plans of the Ottoman capital in the coming years (Pedley, 2012: 29). This first scientific study paved the way towards preliminary small-scale urban interventions in İstanbul at the time. With administrative reforms coming next, the interventions were brought onto a legal ground under the framework of urban planning. One of the most prominent reasons for this was the desire of Ottoman bureaucrats to raise İstanbul to the standards it deserved among other European capitals.

Another urban intervention by the political authority on İstanbul was carried out in the early 19th century by Sultan Selim III. Selim III ordered Üsküdar Palace to be demolished in 1794 and then established the grid-plan districts of Selimiye and İhsaniye on the palace's land, whereas his successor Sultan Mahmud II sacrificed many structures on the Golden Horn coastline to establish filatures and fez manufacturers (İgüs and İsmailoğlu, 2016: 121).

Development activities in the Ottoman Empire were brought under regulation in the 19th century. *Ebniye-i Hassa* (Directorate of State Buildings) was established in 1831, later followed by *Nâfia Nezâreti* (Ministry of Development).

Fig. 2: A Picture Depicting the Plan of İstanbul by Helmuth Von Moltke.
Source: (https://www.researchgate.net)

Issues related with urban renewal in that period were carried out with the help of foreign consultants. For this, a scaled map of İstanbul was drawn. Helmuth Von Moltke, a famous consultant to Mahmud II, was assigned in 1837 by the Prussian government to draw a map of İstanbul and then prepare a development plan for urban improvement.

Moltke not only fortified citadels on both coasts of İstanbul but also created a detailed map of the city, wandering street by street between 1837 and 1839. In his memoirs, Moltke indicated the following about the reaction from townsfolk towards foreign experts: "Turks used to say 'map' when they saw me. While I wandered through neighborhoods, the people would say 'What is this guy even drawing, walking every street? Or is it that the Sultan does not know İstanbul well and wants to be taught by him?'" (Ergin, 1995: 1244). As can be understood from the statements, public expectation from foreign experts was off the point. Moltke also expressed that "One cannot help but ask what could happen in İstanbul if a decent government and an industrious folk existed here" on his impressions of improvements in İstanbul. In addition, Moltke remarked on his detailed map by noting that "It was great pleasure for me to guide my friends whilst riding and boating in this beautiful region where I learnt every corner thanks to mapping" (Moltke, 1969: 7–115).

Moltke also presented significant development decisions in addition to demonstrating the status of the province with his map scaled to 1/25,000, prepared between 1836 and 1837. It is stated that a summary of these decisions is found in the *İlmühaber* (Certificate Registry Book) dated 1839. The most important influence of Moltke was that he pioneered the publishing of *Ebniye Nizamnamesi* (Development Regulation) dated 1848 (Tekeli, 1985: 885). With recommendations prepared by Moltke, solutions were brought to flaws such as narrow and winding roads within the walls of İstanbul, dead-ends, lack of open space, wooden structures, and building inspections. According to this plan, roads were to be widened by certain ratios, houses to be built with stone and no more than 3 floors. Further, it was also recommended to widen the roads in the districts of Divanyolu, Beyazıt, Aksaray, Silivrikapı, and Edirnekapı. Among Moltke's recommendations in the plan, widening roads and constructing squares were the most significant elements that should have existed in such a European city as İstanbul. However, as the first official planning initiative for İstanbul, the plan could not be implemented (Kuban, 2000: 351-352).

Primary goal for Moltke was to create a convenient and uninterrupted transportation network in the historical peninsula of İstanbul where intensive administrative and commercial activities were conducted. In the meantime, houses were planned to be gradually converted from wood to stone or brick in order to prevent fires. On how fires turned into disasters due to all houses in İstanbul being made of wood at the time, Moltke stated that:

> "Houses in İstanbul all are made of wood, even the huge palaces of the Sultan are nothing more than giant huts. It would not be difficult to imagine fires raging through thousands of adjacent and irregular houses, almost made of matches, compressed within only a square mile. In the district of Beyoğlu, large stone houses with shuttered windows began to be erected. But these, too, often fell victim to fires as temperatures only caused by such a sea of flames would be enough to ignite them from within. Extinguishing a fire here is almost unthinkable. [Because] houses are narrow and tall, stairs thin and decayed. A single shout alarm of "fire!" in the dead of night can jolt people out of bed." (Moltke, 1969: 77-78)

Some of the main arterial roads of the present day in the historical peninsula were based on Moltke's plan. The main purpose of this plan was to widen the old Byzantine roads to provide transportation for major neighborhoods and gates. It was envisaged to open new roads with certain dimensions and remove dead ends for urban transportation to flow conveniently. Planting trees on both sides of the new roads and building squares in front of important structures such as mosques were also in the plan. Order in street plans was one of the crucial aspects. Street alignment would strictly determine the border with plots of land.

Except for the mosques, the locations of which coincide with new roads, other public buildings and fountains were to be relocated to better spots. Moreover, coasts of the Golden Horn were to be cleaned and re-organized to improve the scenery. Image was a substantial element of this plan, too. Accordingly, the capital was to transform into a Western and modern city befitting to the Tanzimat philosophy (Çelik, 1998: 84–86). Making this a reality would also mean a dream come true in urban image for the Ottoman elite.

3 New Urban Developments with the Arrival of Tanzimat

With the declaration of Tanzimat, a process of centralization began in the Ottoman Empire, leading to faster modernization in bureaucracy and social life. Reflecting development reforms on Ottoman local administrations resulted in urbanization. Urban development activities used to be conducted by the local administration of states or provinces before Tanzimat, whereas after the declaration of Tanzimat they came under the control of the central government. Creation of municipalities and centralization of public authorities that started with the reform also brought informed interventions by the state to urban spaces of İstanbul. In that period, legislative regulations necessary for urban planning and regulation were published swiftly (İgüs and İsmailoğlu, 2016: 122).

There were critical reasons for urban plan and development of İstanbul during the Tanzimat period. These were as follows: contemporaneous efforts by the Ottoman bureaucratic elite to modernize the city, as befits the Empire, pressure from the residing Europeans, impact of growing population, administrative changes in executive organs, demand for new structures, recent increases in the number of large-scale fires, and attempts to integrate the city to international trade and to establish new industries (Uluengin and Turan, 2005: 359). In addition, frequent visits to the city by civilian and formal foreign guests due to increasing foreign affairs as well as integrating the developing transportation network to other countries can be listed among other reasons.

İlhan Tekeli identifies five main problems that had to be solved in Ottoman provinces, especially in İstanbul, in the 19th century. The first was that new institutions established with growing economic relations and the consequent administrative reforms needed to be restructured in the city center. The second was the latest differentiation within housing zones brought by the new social stratification. The third problem was the dramatic increase in migration and therefore population due to major losses of territory leading to opening new housing zones, the fourth was the attempts to rapidly establish infrastructure and urban services such as roads and mass transportation dearly needed by the resultant urban context, and the fifth was the colossal devastation by fires as most houses were made of wood. For the reasons above, urban regulation in İstanbul became mandatory (Tekeli, 2013: 41).

Administration reform that came with the Tanzimat also brought serious demand for regulation. However, administrative approaches that controlled development in İstanbul in this period failed to display consistency. Developments indicated to the new lifestyle brought by tendency towards European culture. During this transformation in the urban center, old structures were completely removed and newer forms and structures were built by making room next to older ones. As the structuring process continued, older forms of structures adapted to the new conditions. In that way, the urban center verged on stratifying itself as it expanded larger. At the time, the urban center became stratified as it used to extend from the district of Galata to Beyoğlu, whereas the historical peninsula extended from the district of Bab-ı Ali to Saraçhane (Tekeli, 2013: 43).

Traditional Ottoman administrative system went through a set of changes via Tanzimat reforms while İstanbul was introduced with the European style of municipality and Western city planning principles made the improvement of urban fabric even easier. Bureaucrats serving in major European capitals, who influenced modernization reforms during this period, desired to see Western-style urban spaces in İstanbul. Having previously served as a diplomat in Vienna, Paris, and London, as well as being one of the authors of the Imperial Edict of Reorganization (Tanzimat), Mustafa Reşid Paşa defended that the new architectural and urban plans to be implemented in İstanbul would make roads wider and more orderly. Along with that, employment of experts with 'architectural know-how' was necessary to prevent fires. For this reason, he wanted Turkish students to be sent to Europe for education on Western architectural style and planning (Çelik, 1998: 28-41). That would mean transformation of İstanbul from an old town to a modern city. Fires causing large-scale destruction as a crucial problem became an opportunity for transformation in İstanbul's development. They not only paved the way towards the transformation of the urban fabric as per needs, but also gave rise to the evolution of a new understanding of development (Tekeli, 2013: 48).

During his mission in London as ambassador, Mustafa Reşit Paşa indicated in a formal letter he sent to İstanbul that outbreaks of fire in the Ottoman capital were often published in European newspapers along with irritating criticism. It was defaming for the Empire that precautions were not taken despite buildings in İstanbul being generally made of wood. Mustafa Reşit Paşa submitted a detailed letter containing the criticism from European newspapers about the fires. Listing the reasons for urban regulation in his letter, Mustafa Reşit Paşa pointed out that wooden buildings would burn easily and their durability would be less than fifty years even when protected from fire or built sturdy. He recommended popularizing stone structures, encouraging the community to build with stone in addition to bringing in architects and engineers from Europe as soon as possible due to the lack of experts with architectural know-how so that locations damaged by fire can be re-built with stone materials. Further, he reported the necessity of sending pupils to study in Europe so that experts in architecture could be raised. For necessary works to be implemented in a coordinated manner, he suggested the selection of a pilot area in the historical peninsula. According to this suggestion, buildings and shops would be re-built with stone while streets were planned to be re-organized, and the public would be encouraged to build with European style. The interdiction of re-building with wood in districts devastated by fire and the regulation of streets against future fires were among the most important measures. Stone structures would apparently be in the interest of people, and in that way İstanbul would attain the order and beauty it deserved as the center of the Caliphate, drawing the attention and appreciation of Europeans, as well (Baysun, 1960: 124–127). Among the reasons for recommendations listed by Mustafa Reşit Paşa for urban regulation of İstanbul was the concern of 'what Europeans would think.' For this, the contrast between the natural beauty and urban appearance of İstanbul was considered according to European standards (Yerasimos, 2012: 510–11).

The reason was that houses constituting the urban fabric of İstanbul in this period were of wood, unpainted, adjacently compressed, and irregular. As described, mice and weasels dwelt in between worn buildings with dreary indoor spaces. Yards would always be dim, wet, and wormy while streets were in ugly condition. Carcasses of cats, dogs, and mice piled beneath cracked and broken walls in mud throughout the year; land plots and ruins filled with heaps of trash, which would disgust people passing by. In addition to such a bad scenery within towns in İstanbul, even the busiest *Bab-ı Ali* Road was so narrow that cats would easily jump across from one rooftop to the other, and women's conversations across bay windows could easily be heard below by people passing by (Ergin, 1995: 1025–26). Primary reasons for the urban fabric in İstanbul being wood were attributed to its strength against earthquakes and healthy characteristics. However, the actual reason was that wooden structures could be built easier, faster, and cheaper compared to stone structures, especially considering the construction techniques of the era (Tekeli, 2013: 47).

Some of the poorly developed areas in İstanbul were also around the city walls. In order to prevent construction of buildings adjacent to the walls, regulations were enforced to avert inelegant sceneries. This indeed was embarrassing in the eyes of Western statesmen. Such constructions and anomalous endeavors were witnessed by European Christians and ambassadors, leading to condemnation and humiliation of the Exalted State. During the period, the justification behind the regulation preventing constructions adjacent to the walls was the shame felt against Westerners and various ambassadors knowing the situation (www.obarsiv.com).

In that respect, it was vital for Tanzimat administrators to embellish the city and regulate in accordance with Western standards. Tanzimat administrators indicated that regulations for İstanbul were supposed to bring a modern outlook. It was their idea that if, as the most beautiful city in the world with respect to natural beauties, İstanbul could be adorned by experts, it would doubtlessly be the most beautiful city in Europe. According to Tanzimat administrators, beauty and adornment in urban design meant order and regulation. It was necessary to bring order to the urban fabric in order to keep up with the beauty of European cities, and intensive development activities were initiated for wider roads and proper urban design (Çelik, 1998: 128). Such a transformation in the Ottoman Empire was in fact a reflection of intellectual modernization. The new development legislation published in 1839 was in fact a proclamation of wider roads, open squares, and decent streets to be built in İstanbul.

One of the plans created by emulating large-scale development operations implemented in Europe was prepared by Bekir Paşa during the reign of Sultan Abdülmecid. Bekir Paşa was sent to London for education and assigned as the Minister of the Imperial School of Military Engineering to prepare a plan for İstanbul. In the plan, monumental sacred grounds such as mosques were fundamental. Mosques of Süleymaniye, Şehzade, Beyazıt, Sultan Ahmet, Eminönü Yeni Mosque, Laleli, and Nuruosmaniye Mosques were fundamental in the design as all roads in İstanbul were planned to lead to monumental structures while big mosques lied at the center of wide squares. This was a painstaking operation including the protection of monumental structures in order to "endear and familiarize nationality to the nation, and be revered by other nations" (Engin, 1995: 1245). In addition, it was aimed to build parks and gardens within neighborhoods while making room around important public structures such as fountains, water supplies, shrines, and madrasas. Consequently, the historical fabric of the city would be protected. Moreover, it can be said that this plan was prepared by emulating large-scale development operations in European capitals of the time. This emulation was more similar to radical regulations of Baron Hausmann in Paris. For example, Ahmet Vefik Paşa, after being appointed as the governor of Bursa, tried to implement what he learnt while serving as ambassador in Paris during the 1850s (Tekeli, 2011a: 54–55; Tekeli, 1985: 885–886).

The Development Regulation dated 1848 was later published for the same purpose. According to thirty-three articles in this regulation, roads were to be widened and dead ends would be opened as new buildings were erected. Construction of charity structures such as mosques, shrines, schools, and fountains would be encouraged to be built from stone with certain dimensions as per the width of the street at its location. Front sides of existing mosques and madrasas would be organized and expanded while licenses would not be given to constructions at pier squares, mosque courtyards as well as other squares. New buildings and shops would be constructed at a defined elevation with certain construction materials under a certain plan (Ergin, 1995: 1044–49). Other than this regulation, many others were published between 1848 and 1882. Regulations regarding streets, roads, construction activities, as well as Municipality Law and Development Law were also put into force. There were a set of rules with regard to expropriation, supervision of constructions, licensing buildings, widths of roads and streets, as well as heights of buildings to be constructed. Houses were expropriated by the reason of 'public interest' while enforcing contemporary implementations in common with European urban fabric (Çelik, 1998: 42–44).

A new municipality, comparable to the municipality of Paris, was established, named the Sixth Department of Municipality in İstanbul, with the modernization efforts of the Tanzimat period. The second article of the regulation of this institution specified organization of streets, construction of new pavements, regulation of water and sewerage networks, ensuring cleaning and order, as well as a special allowance for their implementation. All of this was to make İstanbul look more contemporary, clean, and orderly by making artistic additions to the natural beauty of the city of the sultanate (Ergin, 1995: 1601).

He was privileged with the execution of road and construction works at the Sixth Department of Municipality. Being outside the legislation of the Directorate of State Buildings, the Sixth Department began working on widening roads in Beyoğlu and Galata, making them fit for vehicles, lighting, and organization of garbage collection. The municipality drafted a cadastral map of the area for the first time and practiced taxing. One of the most important works was the planning of Karaköy footing of the Galata Bridge. This area was one of the busiest places with a ferry quay for urban transportation, wine merchants, and commercial activities through the port. Some of the unseemly and old buildings and shops here were demolished, roads were widened and regulated. Galata walls were removed with a small-scale urban renewal operation. A new commercial complex was built at the land. Loans were taken from the bankers of Galata for this small-scale urban renewal project. Another significant project in this area was the road that connected *Cadde-i Kebir* (Main Road) to Karaköy docks, which was opened and regulated to accommodate vehicles. Falling within the area of the Mevlevi cemetery, this project was still implemented despite objections from the dervishes (Tekeli, 2013: 52-53).

Enforced in 1863, *Turuk ve Ebniye Nizamnamesi* (Street and Development Regulation) constituted a milestone in the history of urban planning in the Ottoman Empire. With this regulation, a condition was brought to implement all development activities in İstanbul and other provinces within the framework of a certain plan (Tekeli, 2013: 60). In this way, planning activities in the capital were extended to other cities as a theory and urban planning was considered to be not only for İstanbul but also an essential necessity of urbanization (Uluengin and Turan, 2005: 366).

Detailed provisions were introduced with regard to roads via this regulation. According to this, the roads were arranged in four stages and signs were installed indicating how wide each street in the city would be at the beginning. Apart from wider roads, it was also aimed to improve the curved roads by pulling them back while newer buildings were constructed. For this reason, maps of routes were prepared in older and newer areas of the city to be regulated. Within fifteen days after these maps were issued to real estate owners, they would have the right to appeal. When the map was finalized, directional signs were placed per road and new constructions and regulations were executed accordingly. In addition, it was aimed to provide solutions to minor problems of urban infrastructure for the first time. Once again for the first time, infrastructures such as water, sewage, and gas pipes were being mentioned. Costs were billed to the owners while naming and numbering the streets. Regulation of possible locations of fire for more effective restructuring of the residential areas became more important. According to the regulation for fire areas, a map specifying land parcels belonging to each property owner would be created for the calculation of property owners' land parcels by deducting road width margins. Further, measures were introduced to reduce fire hazard associated with structures to be built in accordance with the regulation. For those who contradict the regulation, fines were introduced for the first time. This legislation was in force for about twenty years (Tekeli, 2013: 60-62).

3.1 Regulations Enforced due to Fires

Destruction by fires caused by the wooden texture of houses in İstanbul was on a critical scale. During the 19th century, there were more than a hundred fires in İstanbul and many buildings and social structures were damaged. Apart from the smaller ones, the historical peninsula had burnt down from the Cibali district all the way to the Marmara coast, eight times. The most important networks of İstanbul were severely damaged by these fires. Whenever there was a fire, a number of new administrative practices, laws, by-laws and regulations were taken. After the fires, a number of aids and facilitations were provided in the construction of buildings belonging to the community. Buildings of public interest were restored primarily by the sultan, state administrators, and philanthropists (Ergin, 1995: 1239).

The fire of Aksaray in 1856 and the fire of Hocapaşa in 1865 (the great fire) led to reshaping the historical peninsula. After the Aksaray fire, maps of the burned areas were systematically created and a new city design project was prepared and put into practice. Mustafa Resit Paşa, Grand Vizier of the period, appointed a foreign expert after this fire to modernize the city. Luigi Storari, an Italian expert, primarily engaged in street regulation for the reconstruction of the region. The main roads in Aksaray were expanded and combined with other roads of the city (Çelik, 1998: 45).

With the establishment of *Islahat-ı Turuk Komisyonu* (Commission for Road Improvement) in 1866, a new texture was brought to the city with many new street and road regulations. The devastation brought by the great Hocapaşa fire of the previous year led to the urgent establishment of this commission and with this *Men-i Harik Nizamnamesi* (Regulation for Prevention of Fires) was published to protect the city against fires while accelerating the construction of wider streets and buildings. With the expansion of main arteries in the city, broad open spaces were planned to be opened up around big and symbolic mosques. The Commission, responsible for contemporary planning of the whole city, expanded important links in the historic peninsula and established a regular network of transport. Divanyolu, Beyazıt, Süleymaniye, Hagia Sophia, Sirkeci, and Aksaray were all regulated; wider roads were opened; and some structures were demolished.

The Commission for Road Improvement drafted a map of the fire site and prepared a set of plans in accordance with the legislation by expanding and rectifying roads and parcels. Wastewater drainpipes were installed beneath roads, main roads were arranged with rectangular stones whereas side roads with cobblestone pavement. Stone structures were made compulsory in regulated areas, and taxes on building materials were also removed in order to reduce costs of stone structures (Tekeli, 2013: 77). Within three years between 1868 and 1871, the paved and repaired sidewalks reached about five hundred kilometers in the historical peninsula of İstanbul. For this, a large amount of allowances was allocated. This fund was partly obtained from the state and partly from the sale of lands acquired via allotment (Ergin, 1995: 1355)

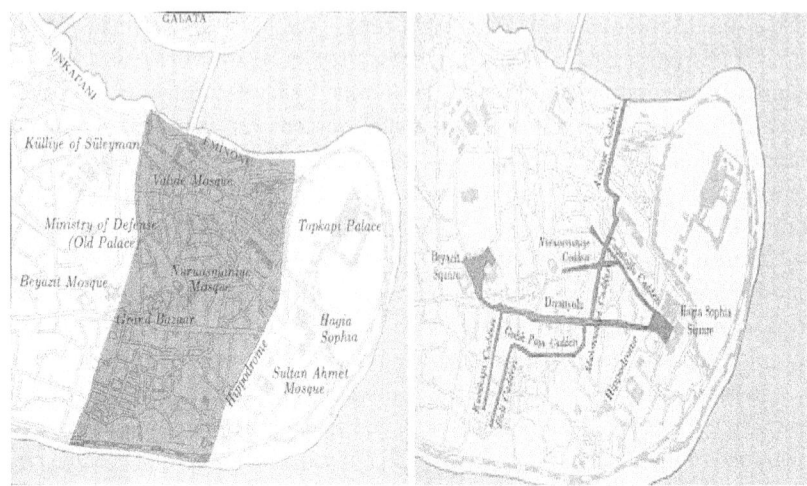

Fig. 3: Hocapaşa Fire Area and Roads Built in the Aftermath. Source: (https://www.arkitera.com)

These modernization practices, carried out by the Commission for Road Improvement in the historical places of İstanbul at neighborhoods affected by fire and at Divanyolu, triggered reaction from the public. For example, some of the unburned buildings after the fire were demolished to widen Divanyolu while others were shrunk for the road. The tombs of Köprülü Mehmet Paşa and his family located on the road plan were moved farther. The public displayed negative reaction to this, which fell on deaf ears, and then development activities continued (Ergin, 1995: 955).

In the historical peninsula, Divanyolu was connected to Beyazıt Square by demolishing a number of irregular building blocks, thus the Divanyolu arrangement project was completed as the most important artery of the city. In its new form, Divanyolu was arranged with sidewalks for the convenience of pedestrians, making it an important artery for easy transportation. The mayor of the period, Server Efendi, reported that the arrangements were performed in accordance with the procedures applied in European cities to leave pleasant impressions on foreign visitors. Appealing to Western visitors, in a way, became a design criterion for the city. The main goal of the Commission for Road Improvement was to provide a modern look to the city, according to Western style and liking. Providing services to neighborhoods damaged by fire, however, came second. The Commission targeted the construction of European arteries instead of reconstruction of damaged areas (Çelik, 1998: 52).

Between 1854 and 1859, before serving as the Mayor of İstanbul, Server Efendi served as ambassador in major cities such as St. Petersburg and Paris, witnessing their transformation on the spot. During the four years between 1865 and 1869 when the Commission for Road Improvement existed, Server Efendi conducted the highest number of urban renewal projects of the 19th century as the Mayor of İstanbul. In this period, İstanbul witnessed the most intense urban renewal process of its history. The commission built main arteries in both sides of the Golden Horn and made room in the peripheries of monumental structures such as Hagia Sophia, Sultan Ahmet, Beyazıt, and Çemberlitaş, completing extensive infrastructure operations. Development activities conducted by the Commission for Road Improvement can still be seen in İstanbul at the present day (Çelik, 1998: 52).

After the fire at İshak Paşa, the area between Hagia Sophia and Sultan Ahmet Mosque was seen fit for the construction of a square. In this area, a part of the Kabasakal neighborhood which was exposed to the fire as well as the entire Hagia Sophia fire area were expropriated. Expropriations were conducted with the understanding that "the Exalted State was not after anyone's wealth or assets except for certain properties" (Ergin, 1995: 1024). There were some reasons for building a square in between the two great sanctuaries. These reasons were listed in the relevant formal letter from the Mayor to the Ministry of Interior. According to this, it was planned to demolish structures such as unseemly houses and cafés built in between two mosques in an irregular and unhealthy way. In this plan, it was aimed to combine the Sultan Ahmet square with the rearranged fire area so that it becomes a part of the Hagia Sophia square. Additionally, it was emphasized to modernize this region by removing the city block which concealed the splendor of Hagia Sophia and Sultan Ahmed mosques and then demolishing the houses located in there, followed by merging the two squares (Akgündüz, et al., 2006: 340–341). The large area that exists today between the two mosques was created by the regulatory efforts of the time.

The dilapidated image of İstanbul, worn out by fires, severely disturbed the sultans. For this reason, the capital was given approval for implementing modern standards. This was clearly stated in a protocol issued by the sultan. The reason was that according to the sultan, "İstanbul is the most honorable and precious point of the world as the center of the caliphate and reign and the prowess of the nation. It does naturally befit the honor of this great city to stand beautiful with development, regulation, and grand works of architecture. Despite the facts, houses and manors in İstanbul have been built from wood without due respect to order or standards. Therefore, since fires in these areas cause dire financial losses, it has been deemed fit to construct buildings from stone, while it will be of great benefit to expand narrow and irregular streets to take measures against possible fires, protecting the public hygiene and health" (Ergin, 1995: 939-40).

3.2 Changing Architectural Style in İstanbul

Administrative reforms towards the end of the Empire were carried out in line with increasing relations with Europe. Architecture and changing urban structure of the Empire were signs of westernization. Seen in military buildings for the first time, Western styles became widespread in palaces and administrative structures which were considered as the center of the state bureaucracy. In the following years, İstanbul was greatly influenced by these Western forms. This change of form in urban spaces, considered to be the symbol of Western progress, was implemented with the initiative from the political circles. The change of development implemented in the urban space since the 19th century, with the conscious will of the political power, revealed the political and cultural atmosphere of the era.

During the Tanzimat period, İstanbul's urban layout manifested itself as an object of the new scientific practices required by modernization. Thus, along with the historical texture of the city, the physical environment also gained a new appearance. This was an effort by the administrators of the period to transform İstanbul into a rather European, which they referred to as 'civilized countries,' and developed city (Akyürek, 2011: 168-169). The Western architectural style of the time was applied to public offices, banks, and other public buildings in İstanbul. Public buildings of the modern lifestyle were constructed in a way that displays features of the Western architectural style (Bozdoğan, 2008: 31).

In this period, not only the public spaces of İstanbul were transformed but also palaces which were both administrative and private spaces for the sultans. The architectural style of the newly built Çırağan, Dolmabahçe, and Yıldız palaces was very different from the old. Further, the Topkapı Palace, the administrative center of the Ottoman Empire for four centuries, also took its share from the spatial change in this period. Contrary to its modest appearance during previous administrations of the rulers, the idea of upgrading to magnificent palaces was formed. It was aimed with newly built palaces to demonstrate that the lost power against the West was finally being revived. Increasing diplomatic relations of the Ottoman Empire and representatives of foreign states were then conducted and welcomed in these new places of power.

These new palaces, which were the administrative centers of the state, followed the latest architectural fashion of Europe with their grandiose interior compositions and exterior façades. This change in architecture was a result of reforms that reflected Western values in the political and cultural arena. This change was not limited to places of power or palaces only. Many public institutions, such as military buildings, also experienced a change. Military and government buildings that were designed with glamorous and great posture began to dominate the city skyline. With the domination of large monumental mosques built after the mid-15th century, a new look began to be created with these new constructions. Built near Hagia Sophia, *Darülfünun* (House of Sciences) was the most eye-catching example in the historical peninsula. The Swiss architect Fossati, who designed many buildings in İstanbul and restored Hagia Sophia, was also the designer of the House of Sciences. This structure had a neo-Renaissance style with a large visual impact on the urban character of İstanbul. However, built during this period, Selimiye Barracks also had the most attractive appearance among other the constructions in Üsküdar. The square-shaped structure became symbolic among military structures (Gül, 2013: 55). The massive barracks reflected the reforms carried out in the army as an example and messenger of the new understanding that was added to the traditional image of the city (Can, 1999: 131).

Relations with the West were reflected in the architectural style as well as in the planning of the newly constructed major military structures. Until the final century of the Ottoman Empire, public and administrative structures were built looking towards Mecca. However, built in 1865 between Süleymaniye Mosque and Beyazıt Mosque in the old palace location, the Serasker building (used as the Rectorate of İstanbul University at present) was not looking towards Mecca. This structure was one of the most important administrative structures of the Ottoman Empire but was not oriented towards Mecca as seen in other religious and administrative structures. This was a clear manifestation of the drift towards the West and an axial shift as a result of Tanzimat reforms in the Ottoman Empire.

4 Abdülhamid II and the Committee of Union and Progress Period

The Ottoman reformists agreed on three main problems of the city: irregular street texture, obsolescence, and disunity. İstanbul was in contrast with other Western cities that were the symbol of modernity and austerity according to the reformist bureaucrats. İstanbul had to be modernized in compliance with the ideal of recovering Empire by implementing European style reforms. For the reformists, this modernization was to bring a beautiful scenery by introducing order to the urban fabric and to create an orderly as well as convenient transport network between various districts of the capital city. The solution to these problems was the organization of street patterns, which meant getting rid of multiple narrow and winding dead-end streets while also building wide, interconnected, and straight arteries. Along with this, arrangements were made via new building legislations to create a uniform type of neighborhood pattern. The arrangements were mainly based on ruined buildings in the coastal neighborhoods. These ruined buildings were demolished and wide open areas were created (Çelik, 1998: 127).

The railway project planned to be built at Sirkeci going along the Marmara coast entered service in 1874. The route of the railway was a controversial issue. About the railway route that would have to pass through the courtyard of Topkapı Palace, the sultan showed his strong commitment by stating that "As long as we have railroad, it can run through my back let alone my backyard" (Gül, 2013:75). Places outside the historical walls, therefore, were connected to the city center. The Development Law dated 1882 used to dictate urban renewal operations during the reign of Abdülhamid II. The most important arrangements affecting physical structure and appearance of the city during Abdulhamid's reign were the construction of ports and stations linking seaports and railroads to the city center (Tekeli, 2013: 88). In 1888, the railway network was extended up to Vienna. With this, a city image that accommodates European lifestyle was created in the sultanate city of İstanbul after a long period of urban renewal. European entrepreneurs held control over economic life through all the planning operations and lastly through the extended railway network. Thus, İstanbul improved its connection to the heart of Europe through new means of transport, and the already existing Westernization process accelerated its pace (Çelik, 1998: 83).

At the end of the 19th century, the Ottoman Empire provided social and welfare services to its subjects as it gradually acquired the quality of being a modern state. Abdülhamid II represented the ultimate heed taken by a sultan for advancement with his achievements in education, health, and welfare. These modernization practices in a sense were acts of "emphasizing the authority of the sultan by making their presence felt without being seen" (Deringil, 2013: 55). The societal welfare institutions and hospitals built by the sultan were presented as the greatest proofs of development and advancement since modernization of medical and health institutions was the most important symbol of progress, science, and technology. By means of such modern institutions, the sultan intended his government to change the image of the Ottoman Empire in addition to the strategy of increasing the visibility of his authority, which was seen as less developed and traditional by European states (Özbek, 2011: 139–198). The objective reflections of political discourses on science and advancement on urban spaces were in the form of *Darülaceze* (House of the Poor), Hamidiye Children's Hospital, watch towers, and other modern schools.

Abdülhamid II had French experts create new plans for the development of İstanbul towards the end of his reign. While serving as ambassador in Paris, Salih Münir Paşa encountered journalists and writers in Europe criticizing urban problems of İstanbul, similar to what M. Reşit Paşa, who was considered to be one of the pioneers of the Tanzimat, encountered as well. He describes as follows how Abdülhamid II summoned him to be tasked with finding a solution for the problem:

"We shall either bear the blame and keep quiet, coming to heel with our compromises, or we shall duly clean, adorn, and let prosper our capital. Only you can perfectly execute this. For you have been living in Europe for a long while, travelling a myriad of places of Europe, visiting decorated cities, you have the finesse in engineering, I shall give you extensive authority and influence, go to France and gather truly acknowledged and well-qualified men, bring them here so we shall have a commission composed of officials as you demand and choose to be under my oversight and your presidency, then we shall get to work." (Tekeli, 1985: 887–888)

Antoine Bouvard, the chief architect of the municipality of Paris, was nominated for this job. Bouvard proposed utopian plans and arrangements for the historic peninsula via photographs and maps of the city before coming to İstanbul. The implementation of such a plan prepared without seeing the city beside the inadequacy of the resources to implement this plan was disapproved. Another architect who prepared a project for İstanbul was Ferdinand Arnodin. Arnodin designed the ring road of İstanbul. Two bridge projects linking the Bosporus and the integrated ring road project were submitted to Abdülhamid but were not initiated in the circumstances of the period. These were challenging mega projects.

Partial planning operations continued during the Second Constitutional Era. Another architect invited to İstanbul was Auric, the chief architect of Lyon. Reports were prepared for a general development in İstanbul as well as the development of fire areas. Auric's report proposed construction of main roads, re-arrangement of fire areas, and construction of some tunnels and suspension bridges (Tekeli, 2013: 118). This proposal was later put into practice by Cemil Topuzlu Paşa, the mayor of İstanbul, for urban development. Initiating the implementation of Auric's plans for İshakpaşa-Sultanahmet fire areas, Cemil Paşa reveals his efforts in the expropriation of the city's fire areas during the preparation of infrastructures:

"Expropriating the lands located between Hagia Sophia and Sultan Ahmet Mosques. There was previously a huge neighborhood that would fill both mosques. One night, the entire neighborhood burnt down. It was even rumored that I did not extinguish the fire deliberately. That's a lie, however. Nevertheless, I'm glad that it burnt down. I revoked all licenses for new construction in there. I expropriated all plots of land where now you see a square including a public bath. I intended to build an asphalt square with a grand monument in the center, just like the 'Palace de la Concorde' in Paris, instead of an architecturally unaesthetic, vile, and shapeless garden. I had projects prepared, listed among famous experts of urban science in Europe. In the meantime, I struggled hard to demolish the public bath. However, despite being led by the former grand vizier Sait Paşa, *Muhafaza-i Asar-ı Atika Cemiyeti* (the Committee for the Preservation of Historic Works) objected to my intention." (Tekeli, 1985: 889–890)

Even during the period of the Union and Progress, as in the Tanzimat and previous periods, fires still played an active role in the urban regulation and in the practice of serial reconstruction of İstanbul. Areas devastated in fires in the historic peninsula were expropriated and wide spaces were created for the construction of modern parks. In particular, buildings near historical and iconic mosques were demolished by expropriation due to fires. Generally, coarse plans made by foreign experts were not put into practice. In spite of that, new development and urban legislations were published.

A need emerged in this period for new municipality building. The municipality requested a new municipality building from the government in 1911. As described at the time, the municipality buildings of developed cities were magnificent and spectacular, but İstanbul city hall, which was the center of the Ottoman reign, was a shameful place in such a state of ruin and decay that repairs and renovations were out of the question. Due to the fact that it was the first place visited by many foreign bureaucrats coming to the city, this awful state of the old city hall was a cause of embarrassment against foreign visitors. The officers in charge of arranging and decorating the city had to work in such a place, creating a bad image in the eyes of foreigners. Therefore, it was first requested from the government to use the post office building as the city hall with its imposing structure in Eminönü, but this was not possible. Afterwards, construction of a new building was offered at a location and in a convenient place in the city, worthy of the glory of the Ottoman Empire. It was desired that this new building be constructed in such a manner worthy of the Ottoman sultanate and the caliphate as in developed cities. This location was the area between Sultan Ahmet Mosque and the Ministry of Justice, namely, the İshak Paşa fire area. This area was partly due to expropriation and was planned to be expanded towards the road at the front. An international competition was organized for the plan and project of the new building to be constructed, titled "International Competition Program for the Procurement of a City Hall Project in İstanbul" (Ergin, 1995: 1527–34). But this endeavor failed and Henri Prost planned the present day metropolitan municipal building in the early Republican period.

6 Conclusion

Significant development activities and planning operations were conducted within the final century of the Ottoman Empire. These activities continued on a busy schedule from the start. In addition to obsolescence, dense buildings, narrow and winding roads, and lack of infrastructure in the urban fabric, large-scale fires causing holocausts within the previous century were among the main reasons for change in especially the urban space. However, the Ottoman quest for fresh glory and wonder against the West as well as the concerns of 'what the West would say' constituted a unique milestone for urban change. Planning efforts that started after the declaration of Tanzimat were made to solve the main problems of the city. The reflections of Westernization led to physical arrangement and transformation, including the spatial structure of the city. Bureaucrats in major capitals of Europe also accelerated the transformation via their wish to see Western regulations in İstanbul. Ottoman bureaucrats desired to utilize from knowledge of the West despite their supremacy. Projection of this in the field of urban management was the transformation of İstanbul into a modern city. Various foreign experts were hired to make İstanbul as modern as European capitals as the center of caliphate and sultanate. Moltke's scale mapping operations, the first planning initiative in the 19th century, served as a basis for plans made in the future. New regulations were tools that helped imitate European style in urban design. It was aimed to gain visibility on urban spaces of İstanbul by bringing together administrative reforms in search of solutions to the problems the state had.

Several legislative arrangements were made for the transformation of İstanbul within the final century of the Ottoman Empire. Fires had a critical role in these legislative arrangements. Devastation due to fire made transformation mandatory by paving the way towards legislative arrangements. Planning and transformation in İstanbul were accelerated by commissions established to introduce modern urban management to the city. Urban transformation was pursued through methods accepted in the Western urban model. During this transformation, a myriad of meticulous rules were brought to the construction of buildings from stone under certain principles, prevention of wooden constructions, arrangement of roads and streets based on specified dimensions, ensuring of fire prevention, and allocation of spaces for wider squares and roads via expropriation of areas that suffered from fires. Western urban forms were reflected by the use of new architectural methods during the physical transformation of the city. It was, therefore, aimed to create an urban image in the sultanate city of İstanbul that resembles Western architectural style, while all planning and arrangement activities were considered as indicators of modernity and advancement.

References

Akgündüz, A., Öztürk, S., Baş, Y. (2006). *Kiliseden Müzeye Ayasofya Camii (Hagia Sophia Mosque from Church to Museum)*, İstanbul: Osmanlı Araştırmaları Vakfı Yayını.

Akyürek, G. (2011). *Bilgiyi Yeniden İnşa Etmek Tanzimat Döneminde Mimarlık Bilgi ve İktidar (Reconstructing Knowledge: Architecture, Knowledge, and Power during the Tanzimat Period)*, İstanbul: Tarih Vakfı Yurt Yayınları.

Baysun, M. C. (1960). *Mustafa Reşit Paşa'nın Siyasi Yazıları (Political Letters of Mustafa Reşid Paşa)*, Tarih Dergisi, 11 (15), pp. 121–142.

Bozdoğan, S. (2008). *Modernizm ve Ulusun İnşası, Erken Cumhuriyet Türkiye'sinde Mimari Kültür (Modernism and Nation Building: Turkish Architectural Culture in the Early Republic)*, (Trans.) Birkan T., İstanbul: Metis Yayınları.

Can, C. (1999). "Tanzimat ve Mimarlık", *Osmanlı Mimarlığının 7 Yüzyılı Uluslarüstü Bir Miras (Seven Centuries of Ottoman Architecture: A Supranational Heritage)*, (Eds.) Akın N., Batur A., Batur S., İstanbul: Yem Yayın, pp. 130–136.

Çelik, Z. (1998). *19. Yüzyılda Osmanlı Başkenti Değişen İstanbul (The Remaking of Istanbul: Portrait of an Ottoman City in the Nineteenth Century)*, (Trans.) Deringil S., İstanbul: Tarih Vakfı Yurt Yayınları.

Deringil, S. (2013). *Simgeden Millete II. Abdülhamid'den Mustafa Kemal'e Devlet ve Millet (From Symbol to Nation: State and Nation from Abdülhamid II to Mustafa Kemal)*, İstanbul: İletişim Yayınları.

Ergin, O. N. (1995). *Mecelle-i Umur-ı Belediyye (Ottoman Code of Municipalities)*, 2–4, İstanbul: İstanbul Büyükşehir Belediyesi Yayınları.

Gül, M. (2013). *Modern İstanbul'un Doğuşu - Bir Kentin Dönüşümü ve Modernizasyonu (The Emergence of Modern Istanbul: Transformation and Modernization of a City)*, (Trans.) Helvacıoğlu, B., İstanbul: Sel Yayınları.

http://www.obarsiv.com/vct_0506_ugurtanyeli.html, (10.05.2018).

https://İstanbul-constantinople.culturalspot.org/asset-viewer/i%CC%87stanbul-%C5%9Fehir-plan%C4%B1-ve-semtleri/hwF86T3hRkwFoQ, (20.05.2018).

https://www.arkitera.com/galeri/detay/53232/5, (30.05.2018).

https://www.researchgate.net/figure/The-Moltke-Plan-of-İstanbul-Helmuth-von-Moltke-1839_fig1_276158909, (20.05.2018).

İgüs, E., İsmailoğlu, H. (2016). "Osmanlı Kenti İstanbul'u Yıkamak ve Yeniden Yapmak Paradoksu: Menderes Yıkımları", *Osmanlı İstanbuhe Ottoman İstanbul)*, (Eds.) Emecen, F. M., Akyıldız, A., Gürkan, E. S., İstanbul: İstanbul Büyükşehir Belediyesi Yayınları, pp. 115–158.

Kuban, D. (2000). *İstanbul Bir Kent Tarihi (İstanbul: An Urban History)*, İstanbul: Tarih Vakfı Yurt Yayınları.

Le, C. (2011). *Şark Seyahati İstanbul 1911 (Journey to the East: İstanbul 1911)*, (Trans.) Tümertekin, A., İstanbul: Türkiye İş Bankası Kültür Yayınları.

Moltke, H. V. (1969). *Türkiye Mektupları (Letters from Turkey)*, (Trans.) Örs H., İstanbul: Remzi Kitapevi.

Özbek, N. (2011). *Osmanlı İmparatorluğu'nda Sosyal Devlet (Social State in the Ottoman Empire)*, İstanbul: İletişim Yayınları.

Pedley, M. (2012). "Enlightenment Cartography at the Sublime Porte: François Kauffer and the Survey of Constantinople", *The Journal of Ottoman Studies*, 39, pp. 29–53.

Tekeli, İ. (1985). "Tanzimat'tan Cumhuriyet'e Kentsel Dönüşüm", *Tanzimat'tan Cumhuriyet'e Türkiye Ansiklopedisi (Encyclopedia of Turkey from Tanzimat to the Republican Era)*, (Ed.) Belge, M., İstanbul: İletişim Yayınları, pp. 878–890.

Tekeli, İ. (2011a). *Türkiye'nin Kent Planlama ve Kent Araştırmaları Tarihi Yazıları (Letters on the History of Urban Planning and Urban Research in Turkey)*, İstanbul: Tarih Vakfı Yurt Yayınları.

Tekeli, İ. (2011b). *Tasarım, Mimarlık ve Mimarlar (Design, Architecture, and Architects)*, İstanbul: Tarih Yurt Vakfı Yayınları.

Tekeli, İ. (2013). *İstanbul'un Planlanmasının ve Gelişmesinin Öyküsü (The Story of Planning and Development of İstanbul)*, İstanbul: Tarih Yurt Vakfı Yayınları.

Uluengin, M. B., Turan, Ö. (2005). "İmparatorluğun İhtişam Arayışından Cumhuriyet'in Radikal Modernleşme Projesine: Türkiye'de Kentsel Planlamanın İlk Yüz Yılı", *Türkiye Araştırmaları Literatür Dergisi (Literature Journal of Studies on Turkey)*, 3, (6), pp. 353–436.

Yerasimos, S. (2012). "Tanzimat'ın Kent Reformları Üzerine", *Tanzimat Değişim Sürecinde Osmanlı İmparatorluğu (Ottoman Empire during the Tanzimat Reforms)*, (Ed.) İnalcık, H., Seyitdanlıoğlu, M., İstanbul: Türkiye İş Bankası Yayınları, pp. 505–525.

www.ingramcontent.com/pod-product-compliance
Ingram Content Group UK Ltd.
Pitfield, Milton Keynes, MK11 3LW, UK
UKHW021842210426
5322IPUK00022B/411